Eureka Math™
A Story of Units

Common Core, Inc. (commoncore.org) is a non-profit organization formed in 2007 to advocate for a content-rich liberal arts education in America's K-12 schools. To improve education in America, we create curriculum materials, conduct professional development, and also promote programs, policies, and initiatives at the local, state, and federal levels that provide students with challenging, rigorous instruction in the full range of liberal arts and sciences.

Common Core, Inc. is not affiliated with the Common Core State Standards Initiative.

Special thanks go to the Gordan A. Cain Center and to the Department of Mathematics at Louisiana State University for their support in the development of *Eureka Math*.

Published by Common Core

Common Core
1016 16th Street NW, 7th Floor
Washington, DC 20036
Phone 202.223.1854
Web commoncore.org
Email info@commoncore.org

Printed in the U.S.A.
This book may be purchased from the publisher at commoncore.org
10 9 8 7 6 5 4 3 2 1

ISBN 978-1-63255-000-2

Name _____   Date _____

Color the things that are the same. Color them so that they look like each other.

EUREKA MATH™ | Lesson 1:   Analyze to find two objects that are *exactly the same* or *not exactly the same*.
Date: 6/6/14

3

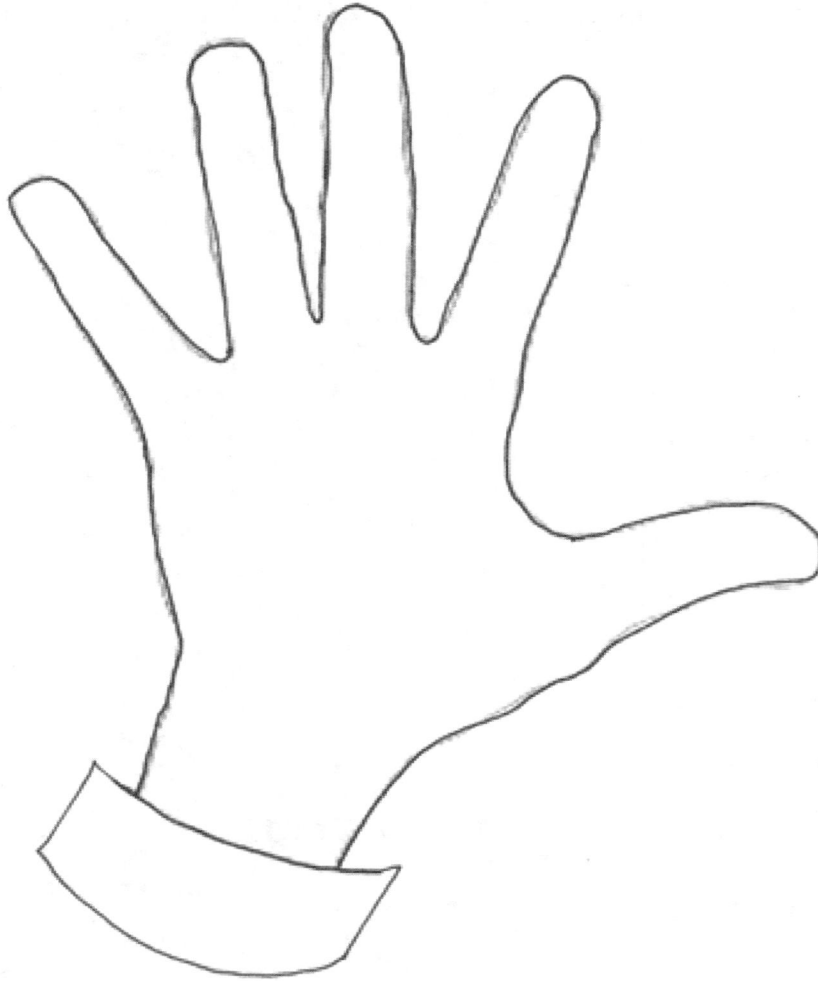

left hand mat

Name _____   Date _____

Are they the same? Circle your answer, and explain it to an adult or friend.

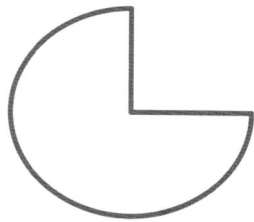

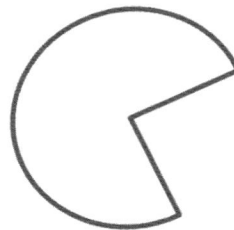

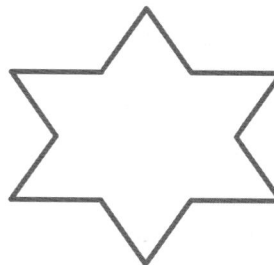

Are these the same?   YES   NO

Are these the same?   YES   NO

Are these the same?   YES   NO

Name _____   Date _____

Draw a line between the objects that have the same pattern. Talk with a neighbor about the objects that match.

EUREKA
MATH™

Lesson 3:   Classify to find two objects that share a visual pattern, color, and use.
Date: 6/6/14

8

Circle the object that would be used together with the object on the left.

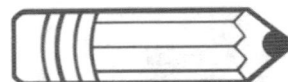

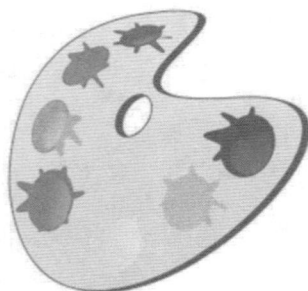

Lesson 3:   Classify to find two objects that share a visual pattern, color, and use.
Date: 6/6/14

9

Name _____ Date _____

Draw a connecting line between shapes with the same pattern.

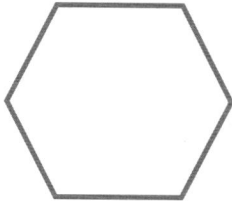

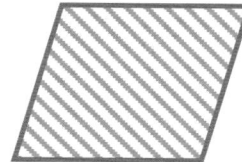

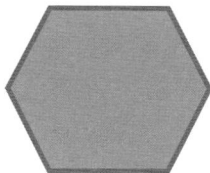

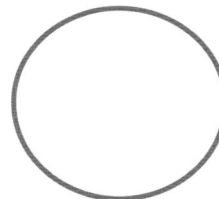

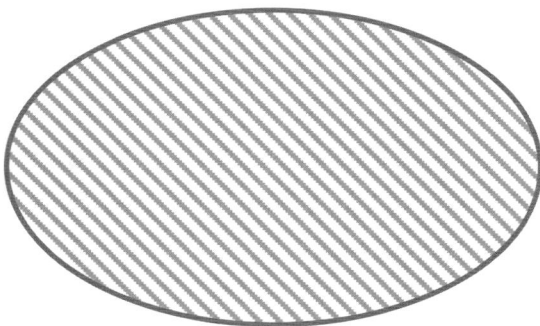

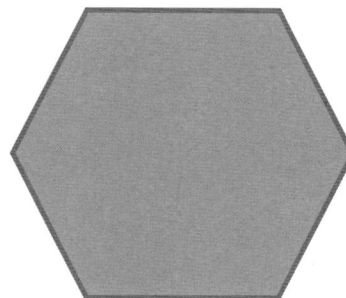

EUREKA MATH Lesson 3: Classify to find two objects that share a visual pattern, color, and use.
Date: 6/6/14

Circle the things that are used together. Explain your choice.

EUREKA
MATH

Lesson 3:
Date:

Classify to find two objects that share a visual pattern, color, and use.
6/12/14

11

Name _____   Date _____

Draw something that you would use with each. Tell why.

Make a picture of 2 things you use together. Tell why.

EUREKA MATH™

Lesson 3:   Classify to find two objects that share a visual pattern, color, and use.
Date: 6/6/14

12

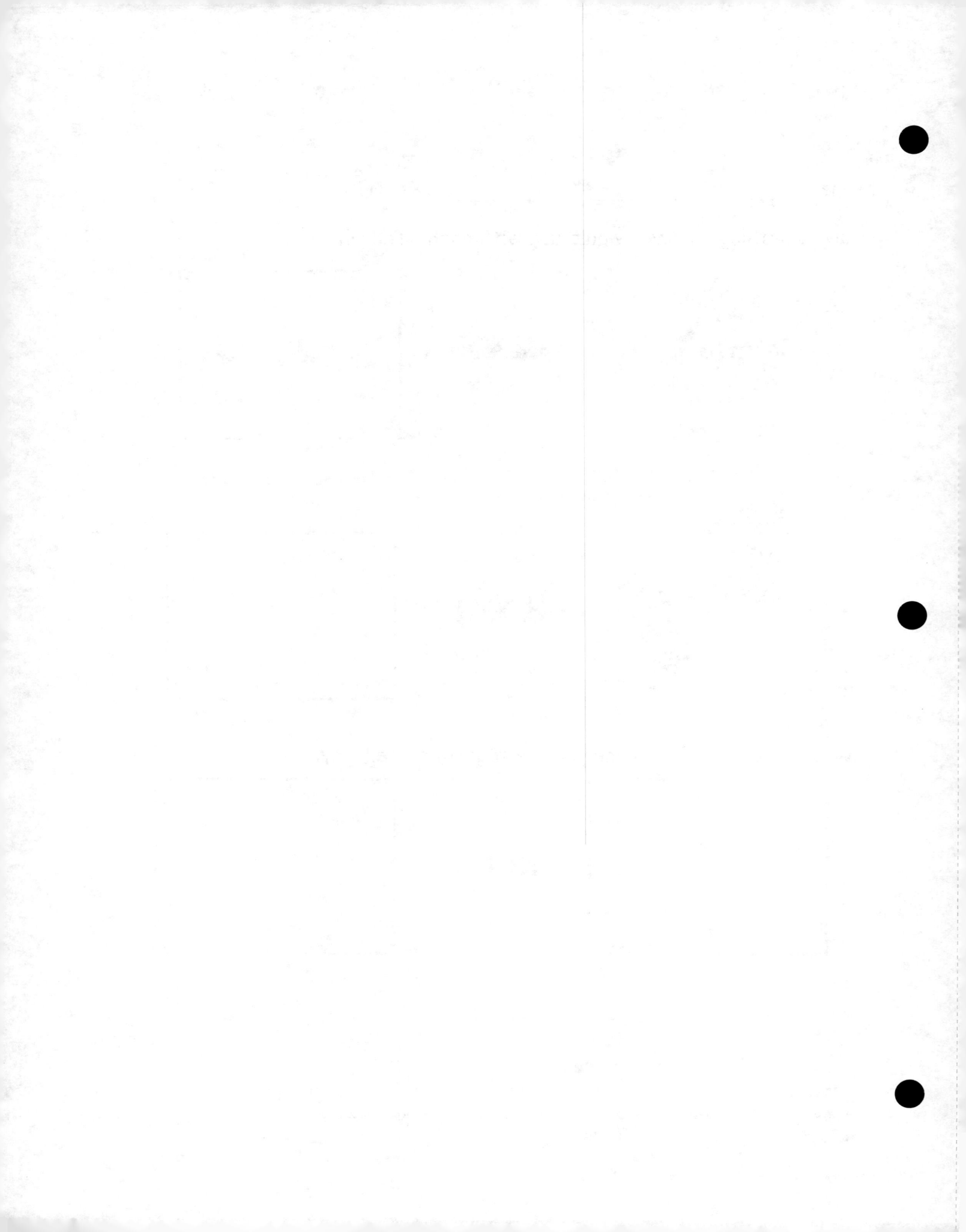

Name _____   Date _____

Use the cutouts. Glue the pictures to show where to keep each thing.

EUREKA
MATH™   Lesson 4:   Classify items into two pre-determined categories.
               Date: 6/6/14

                                                                          13

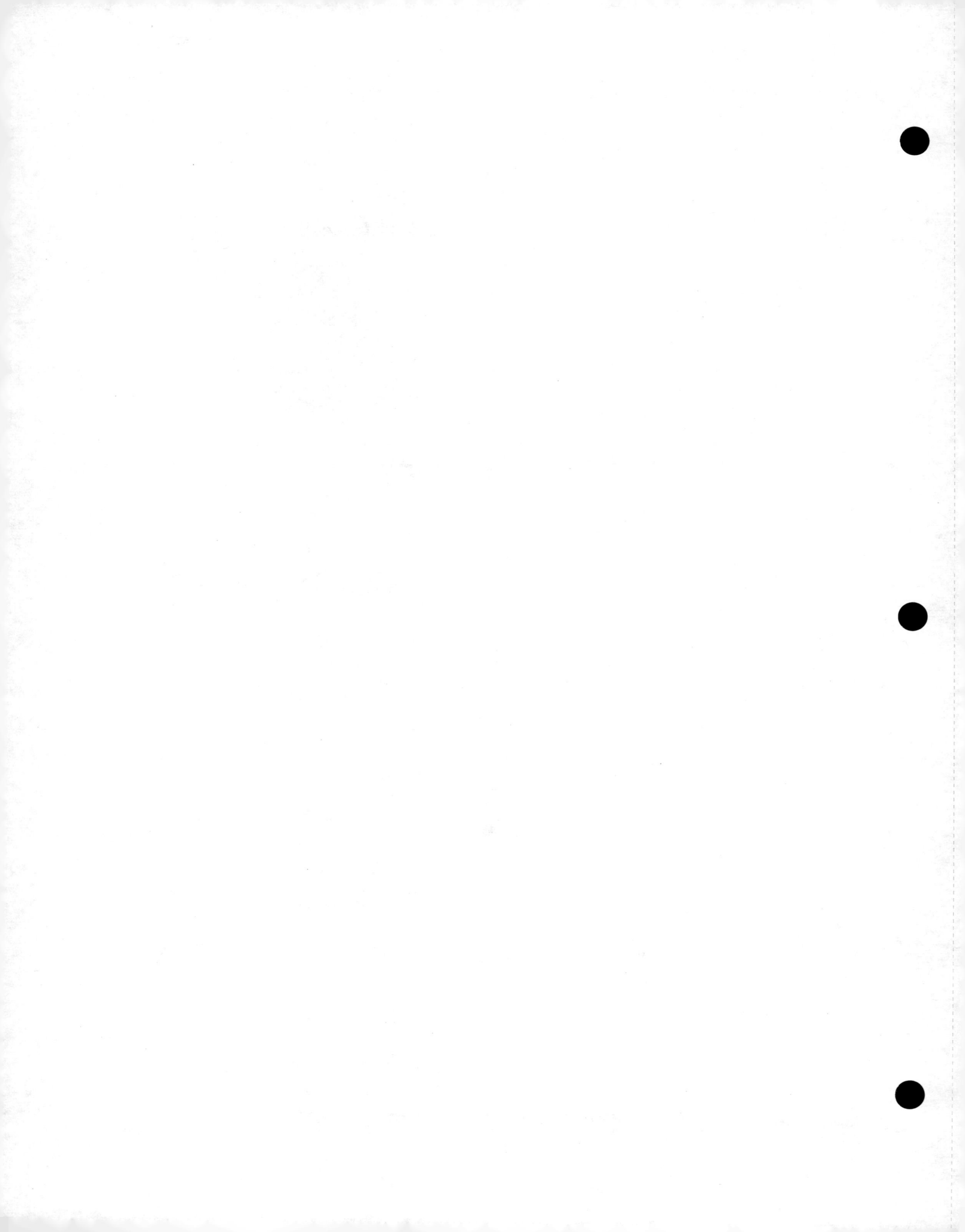

Name _____   Date _____

Cutouts for the Problem Set

EUREKA
MATH

Lesson 4:   Classify items into two pre-determined categories.
Date: 6/6/14

14

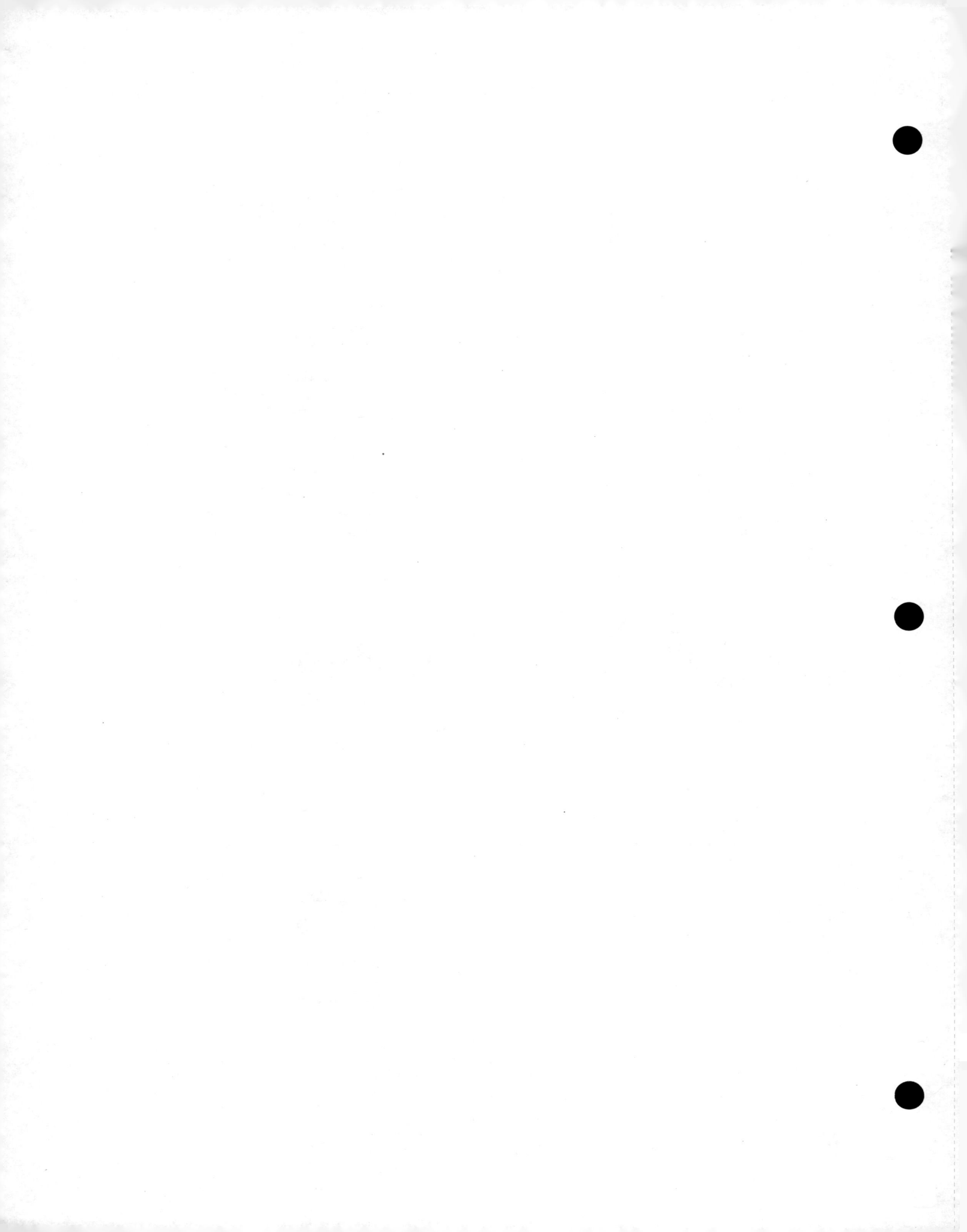

Name _____   Date _____

Circle the animals that belong to one group, and underline the animals that belong to the other group.

What is the same about the animals in each group? (Discuss with a friend.) (Teacher circulates, listening to conversations and making informal assessments.)

Name _____　Date _____

Circle the things that belong to one group, and underline the things that belong to the other group. Tell an adult why the items in each group belong together.

EUREKA
MATH

Lesson 4:　　Classify items into two pre-determined categories.
Date:　　　　6/6/14

16

Name _____   Date _____

Draw a line with your ruler to show where each thing belongs.

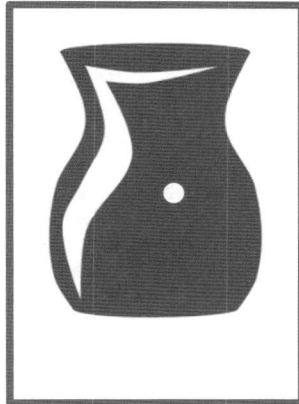

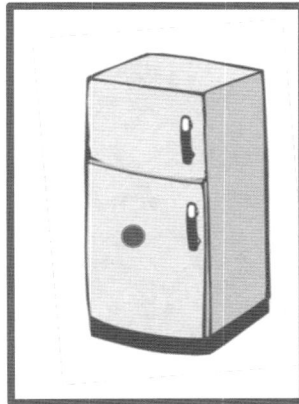

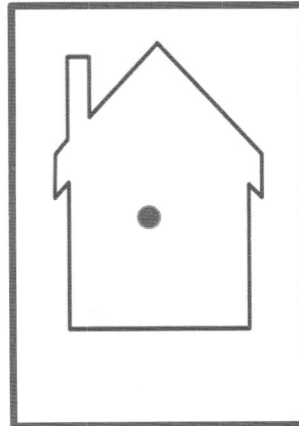

EUREKA MATH

Lesson 5:   Classify items into three categories, determine the count in each, and reason about how the last number named determines the total.

Date: 6/6/14

Name _____ Date _____

Cross out what doesn't belong. How many are left? (Students may cross out more than 1 item in each row. Students explain the group left to a friend or teacher.)

Cross out what doesn't belong. How many are left?

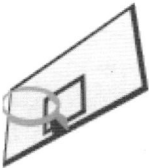

Cross out what doesn't belong. How many are left?

EUREKA MATH

Lesson 5: Classify items into three categories, determine the count in each, and reason about how the last number named determines the total.

Date: 6/6/14

18

Name _____   Date _____

Cut and glue where each belongs. Write how many.

EUREKA
MATH

Lesson 5:   Classify items into three categories, determine the count in each, and
             reason about how the last number named determines the total.
Date: 6/6/14

19

Library

Number:_____

School

Number:_____

Grocery Store

Number:_____

EUREKA MATH   **Lesson 5:**   Classify items into three categories, determine the count in each, and reason about how the last number named determines the total.   20

Date: 6/6/14

The Birthday Cake

The Birthday Cake

birthday cake

EUREKA
MATH

Lesson 5:

Date:

6/6/14

Classify items into three categories, determine the count in each, and
reason about how the last number named determines the total.

21

Name _____   Date _____

Look at the shelf. Color the things in groups of 2 red. Color the things in groups of 3 blue. Color the things in groups of 4 orange.

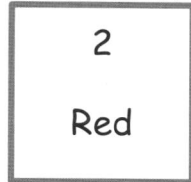

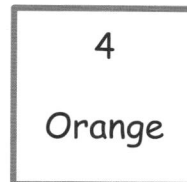

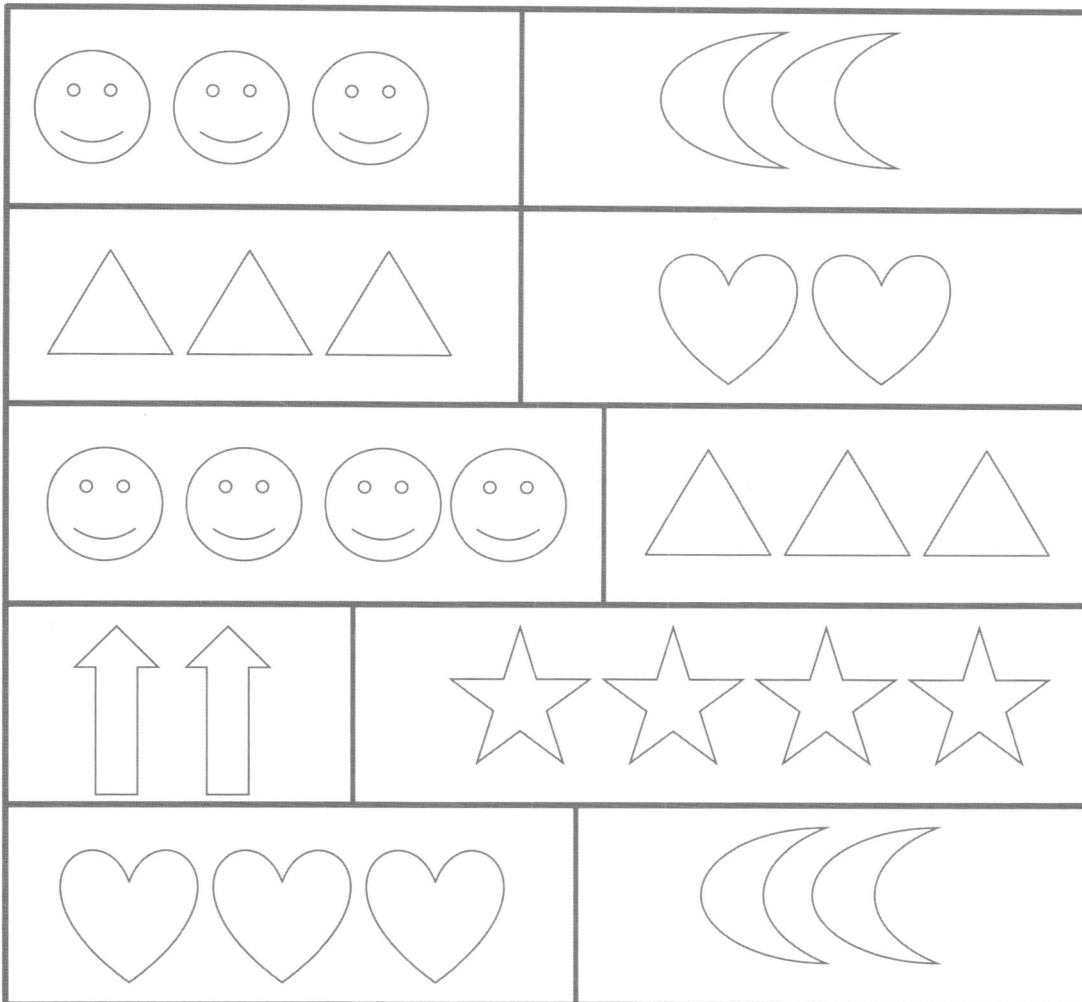

| 2 Red | 3 Blue | 4 Orange |

EUREKA MATH™

Lesson 6:   Sort categories by count. Identify categories with 2, 3, and 4 within a given scenario.
Date: 6/6/14

Name _____   Date _____

Match the groups that have the same number.

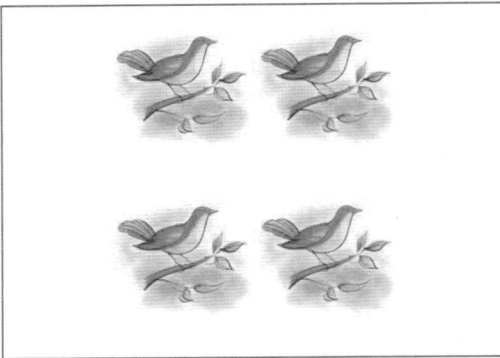

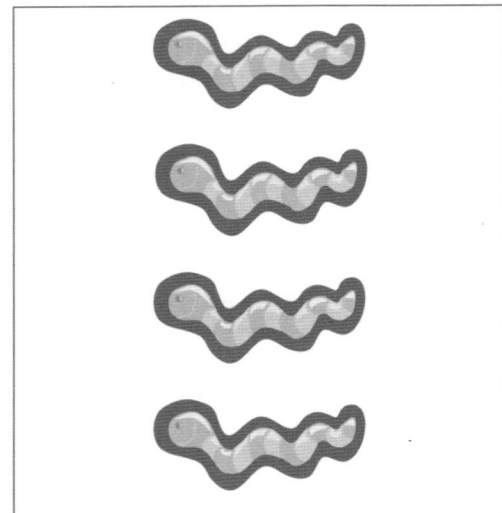

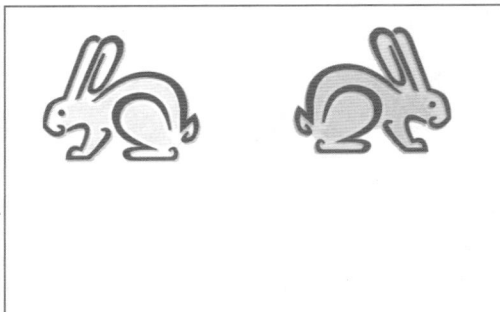

Lesson 6: Sort categories by count. Identify categories with 2, 3, and 4 within a given scenario.
Date: 6/6/14

EUREKA MATH

23

Name _____   Date _____

Draw lines to put the treasures in the boxes.

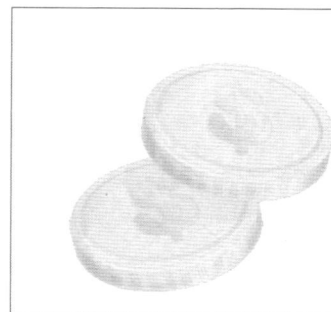

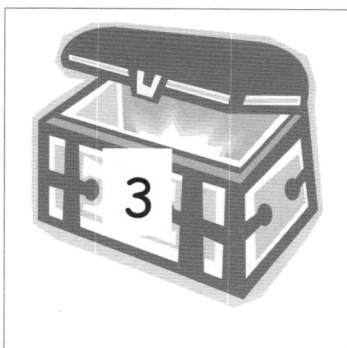

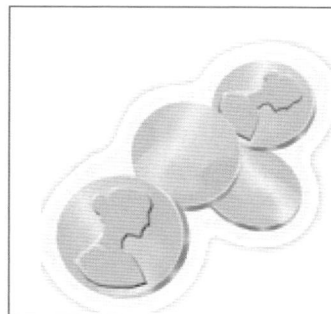

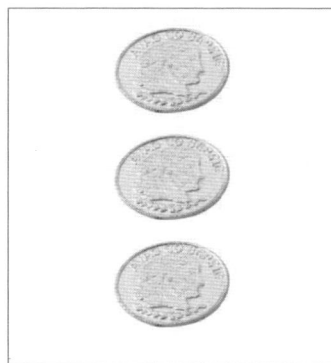

EUREKA
MATH™

Lesson 6:   Sort categories by count. Identify categories with 2, 3, and 4 within a
              given scenario.
Date: 6/6/14

24

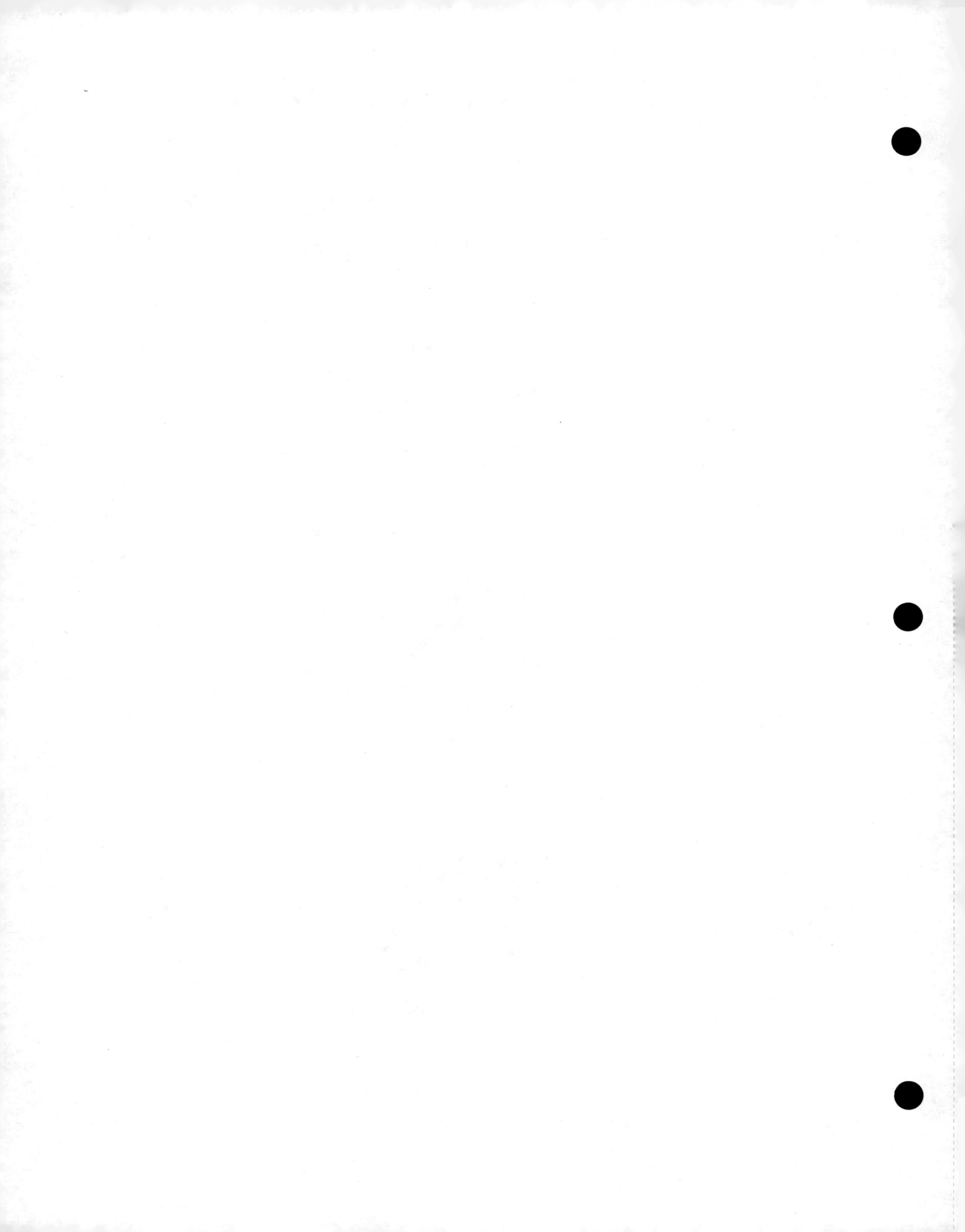

Name _____   Date _____

Color each numeral card as directed. Count the objects in each group.
Then, color the group of objects the same color as the numeral card that
it matches.

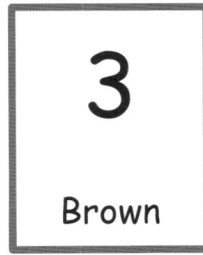

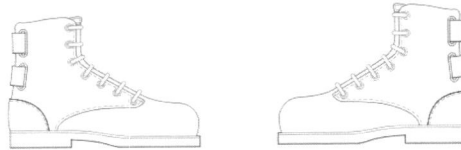

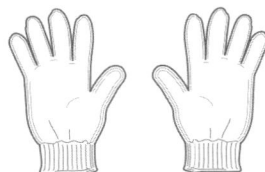

1	2	3	4	5
Black	Blue	Brown	Red	Yellow

EUREKA
MATH™

Lesson 7:   Sort by count in vertical columns and horizontal rows (linear
configurations to 5). Match to numerals on cards.
Date: 6/6/14

25

Name _____   Date _____

Count the shapes. Color in the box that tells how many there are.

3

4

5

EUREKA MATH | Lesson 7: | Sort by count in vertical columns and horizontal rows (linear configurations to 5). Match to numerals on cards.
| Date: | 6/6/14

26

Name _____   Date _____

Color each numeral card as directed. Count the objects in each group.
Then, color the group of objects the same color as the numeral card that
it matches.

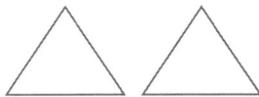

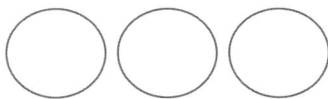

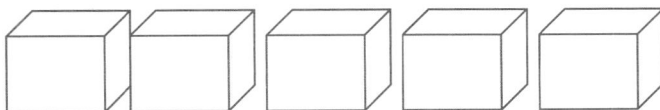

1	2	3	4	5
Black	Blue	Brown	Red	Yellow

EUREKA
MATH™

Lesson 7:   Sort by count in vertical columns and horizontal rows (linear
            configurations to 5). Match to numerals on cards.
Date: 6/6/14

27

Cut out one 5-frame for each student.

5-frames

EUREKA MATH™   Lesson 7:   Sort by count in vertical columns and horizontal rows (linear configurations to 5). Match to numerals on cards.   28

Date: 6/6/14

0	1	2	3
4	5	5	6
7	8	9	10

Note: Only cards 1–5 are used in this lesson. Save the full set for use in future lessons. Consider copying on different color card stock for ease of organization.

5-group cards (numeral side) (Copy double-sided with 5-groups on card stock, and cut.)

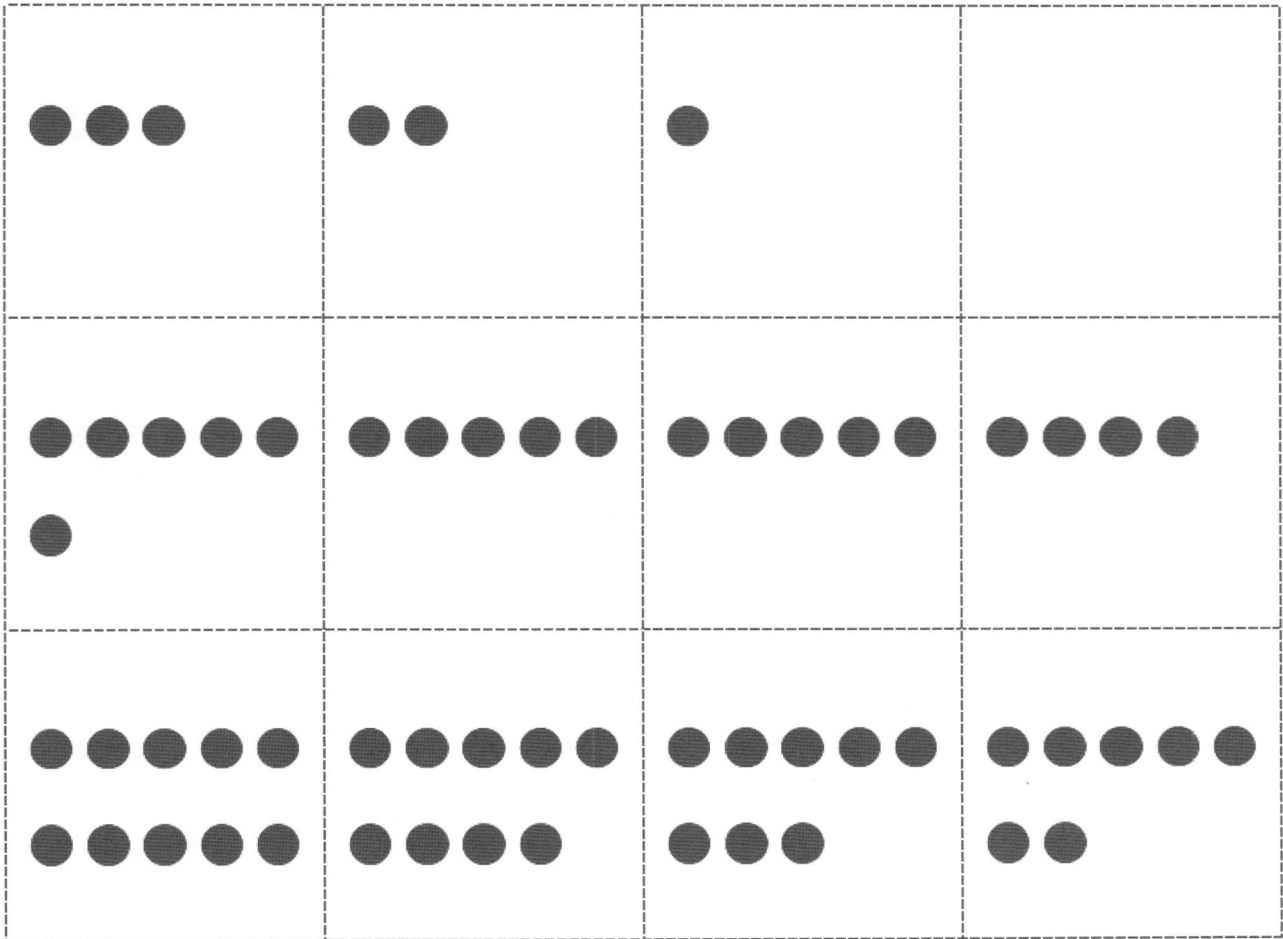

5-group cards (5-group side) (Copy double-sided with numerals on card stock, and cut.)

EUREKA
MATH

Lesson 7:   Sort by count in vertical columns and horizontal rows (linear
configurations to 5). Match to numerals on cards.
Date: 6/6/14

30

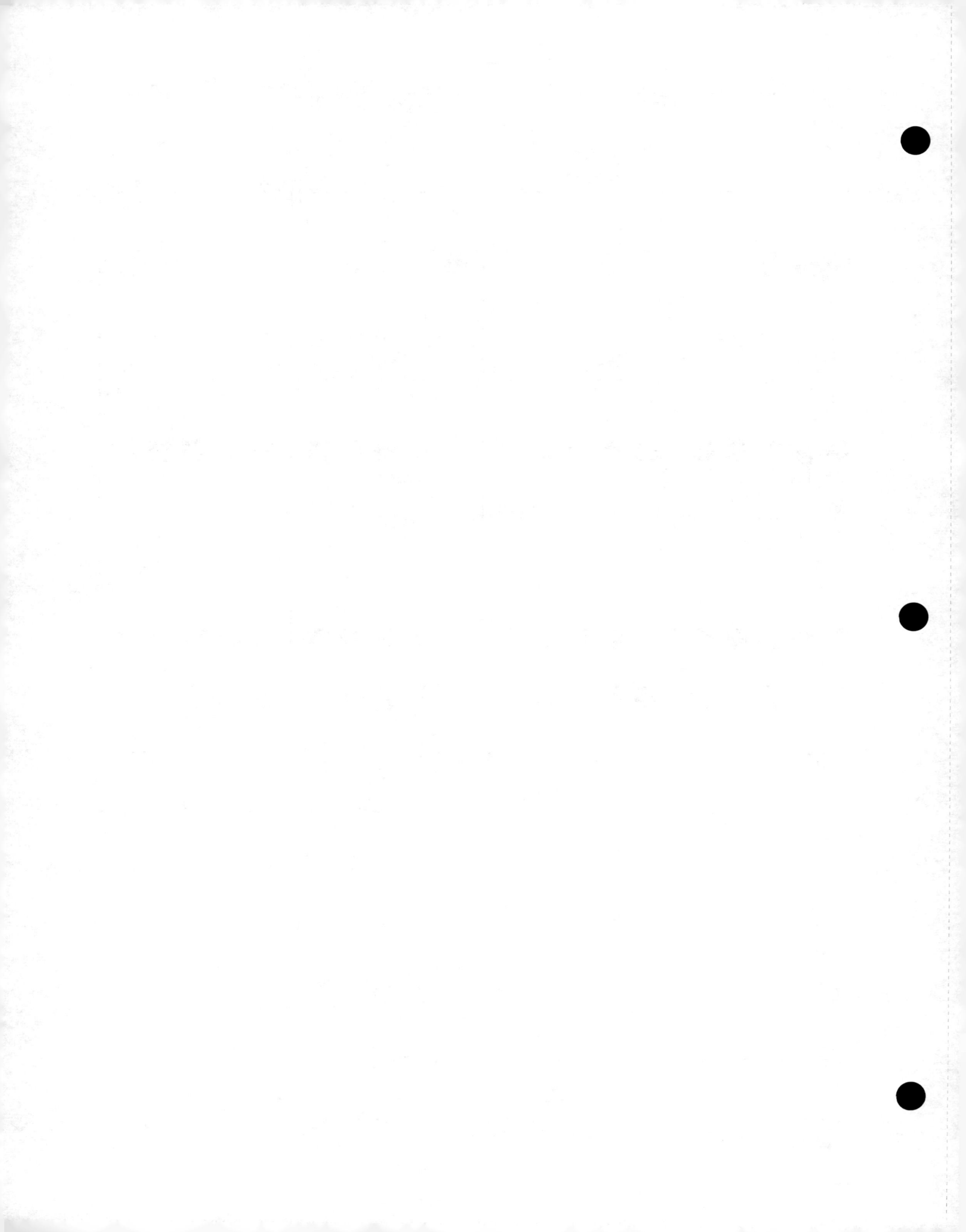

Name _____   Date _____

Count the objects. Circle the correct number.

1   2   3

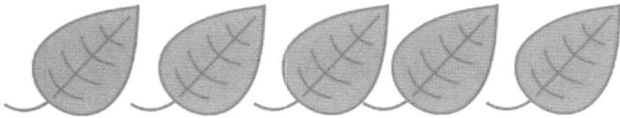

1   2   3

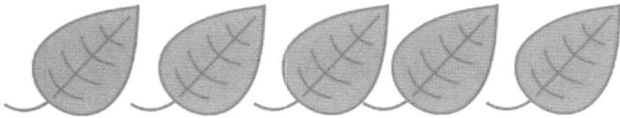

3   4   5

2   3   4

4   3   2

5   4   1

4   3   2

5   4   1

EUREKA MATH

Lesson 8:   Answer *how many* questions to 5 in linear configurations (5-group), with 4 in an array configuration. Compare ways to count five fingers.
Date: 6/6/14

31

Name _____   Date _____

Count. Circle the number that tells how many.

☺ ☺ ☺	1 2 3 4 5
✦ ✦ ✦ ✦	1 2 3 4 5
♡ ♡ ♡ ♡ ♡	1 2 3 4 5
☾ ☾ ☾ ☾	1 2 3 4 5
△ △ △ △ △	1 2 3 4 5
☐ ☐ ☐ ☐ ☐	1 2 3 4 5

EUREKA MATH

Lesson 8:   Answer *how many* questions to 5 in linear configurations (5-group), with 4 in an array configuration. Compare ways to count five fingers.
Date: 6/6/14

32

Name _____   Date _____

Count. Circle the number that tells how many.

●●●●	4	5
(4 dots vertical)	4	5
●●●●●	4	5
(5 dots diagonal)	4	5
(die showing 4)	4	5
(die showing 4) (die showing 1)	4	5
(die showing 5)	4	5

EUREKA MATH

Lesson 8: Answer *how many* questions to 5 in linear configurations (5-group),
with 4 in an array configuration. Compare ways to count five fingers.

Date: 6/6/14

33

© 2014 Common Core, Inc. All rights reserved. commoncore.org

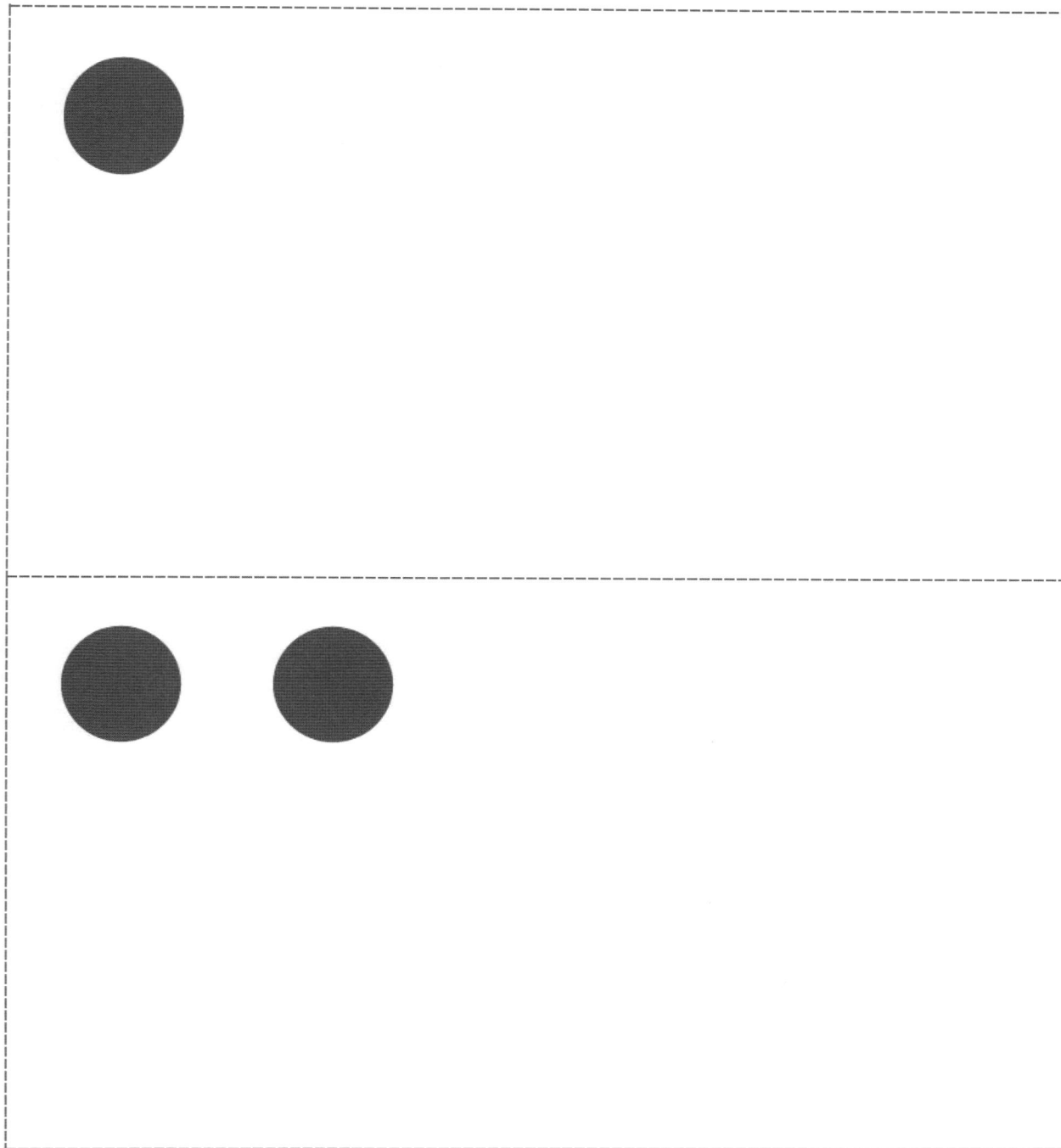

large 5-group cards (Copy on card stock, and cut. Use cards 1–5 in today's Fluency Practice. Save full set.)

EUREKA MATH

Lesson 8: Answer *how many* questions to 5 in linear configurations (5-group), with 4 in an array configuration. Compare ways to count five fingers.
Date: 6/6/14

34

large 5-group cards (Copy on card stock, and cut. Use cards 1–5 in today's Fluency Practice. Save full set.)

EUREKA MATH™

Lesson 8: Answer *how many* questions to 5 in linear configurations (5-group), with 4 in an array configuration. Compare ways to count five fingers.

Date: 6/6/14

35

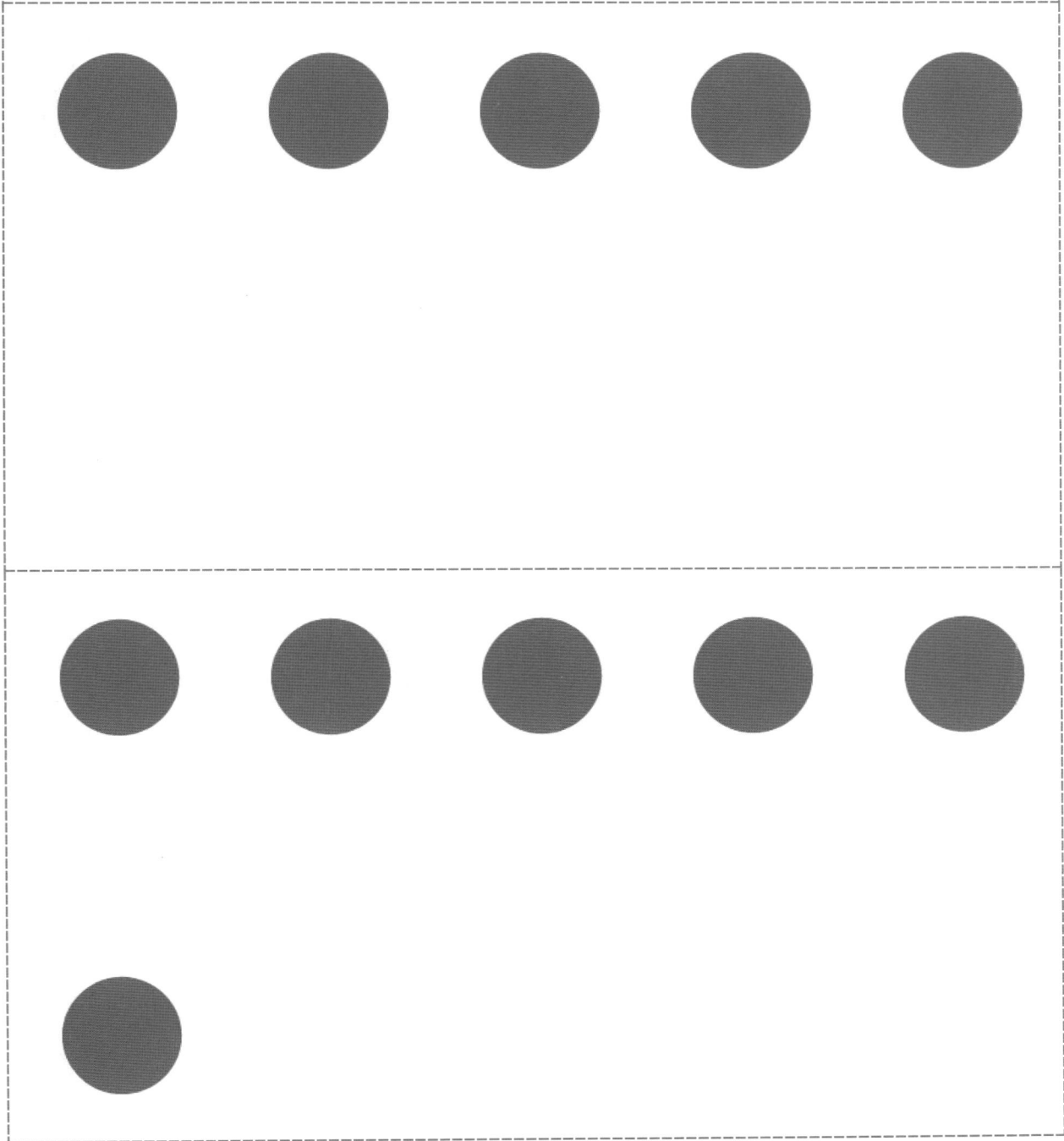

large 5-group cards (Copy on card stock, and cut. Use cards 1–5 in today's Fluency Practice. Save full set.)

EUREKA
MATH

Lesson 8:   Answer *how many* questions to 5 in linear configurations (5-group),
             with 4 in an array configuration. Compare ways to count five fingers.
Date: 6/6/14

36

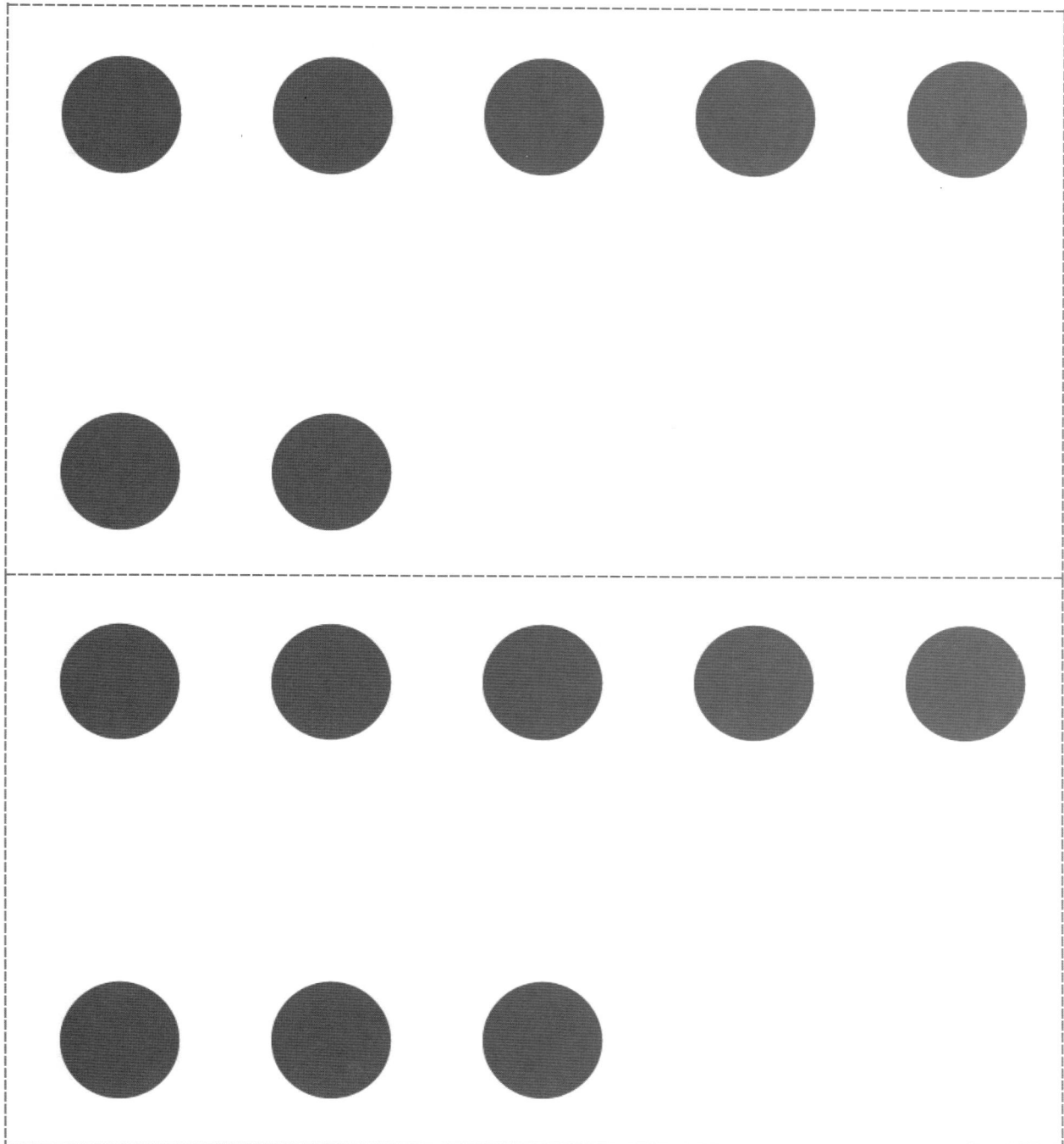

large 5-group cards (Copy on card stock, and cut. Use cards 1–5 in today's Fluency Practice. Save full set.)

EUREKA MATH

Lesson 8:   Answer *how many* questions to 5 in linear configurations (5-group),
              with 4 in an array configuration. Compare ways to count five fingers.
Date: 6/6/14

37

© 2014 Common Core, Inc. All rights reserved. commoncore.org

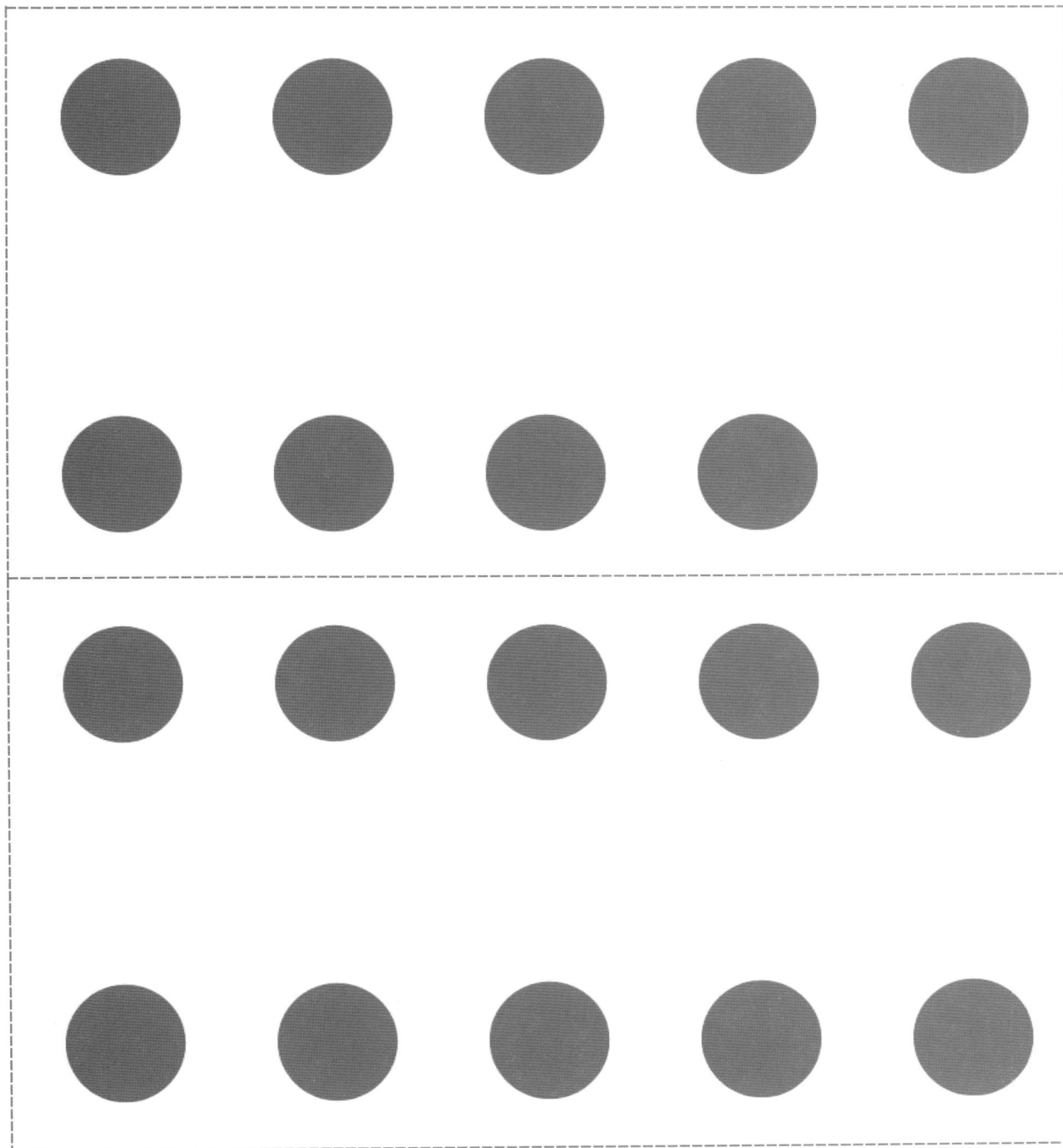

large 5-group cards (Copy on card stock, and cut. Use cards 1–5 in today's Fluency Practice. Save full set.)

EUREKA
MATH™

Lesson 8:   Answer *how many* questions to 5 in linear configurations (5-group),
with 4 in an array configuration. Compare ways to count five fingers.
Date: 6/6/14

38

© 2014 Common Core, Inc. All rights reserved. **commoncore.org**

Name _____   Date _____

Count the dots, and circle the correct number. Color the same number of dots on the right as the gray ones on the left to show the hidden partners.

3   4   5

   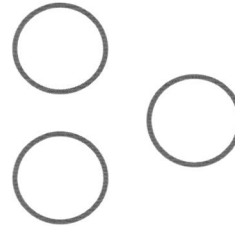

3   4   5

   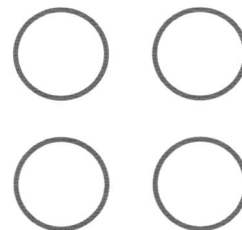

3   4   5

   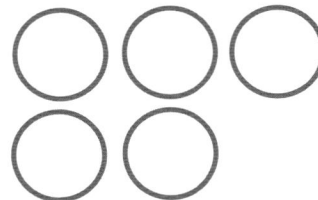

3   4   5

EUREKA MATH™   **Lesson 9:**   Within linear and array dot configurations of numbers 3, 4, and 5,   39
                 find *hidden partners*.
                 Date: 6/6/14

Name _____   Date _____

Circle 3 to show the hidden partners.

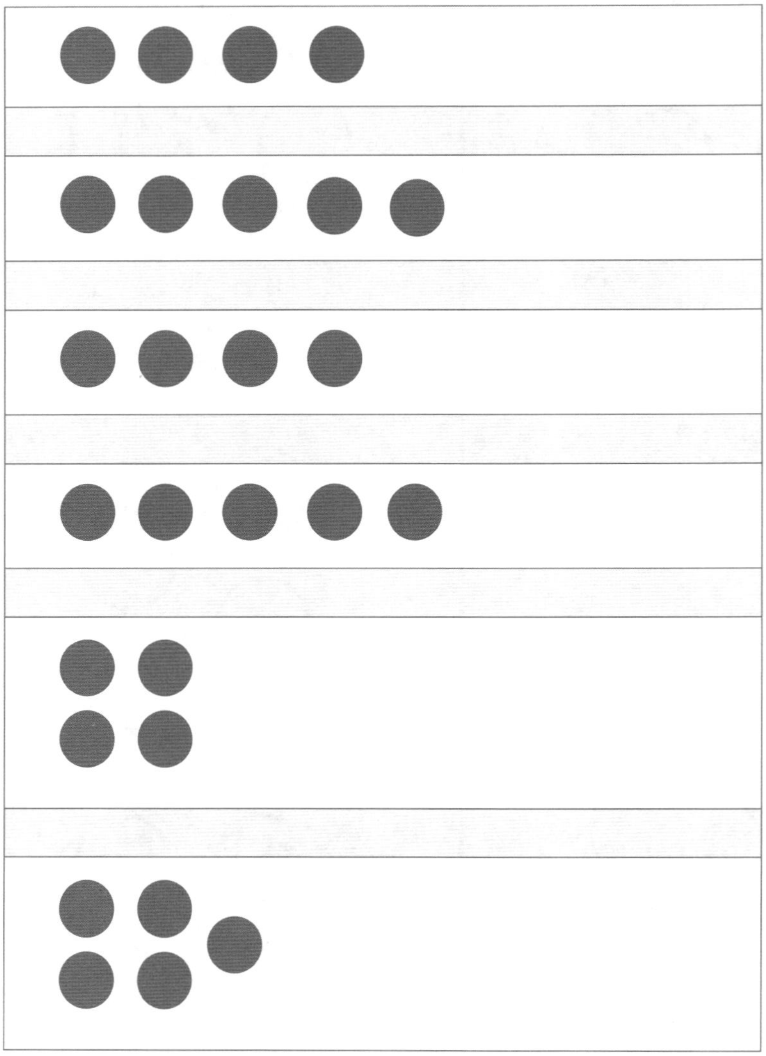

EUREKA MATH

Lesson 9:   Within linear and array dot configurations of numbers 3, 4, and 5,
              find *hidden partners*.
Date: 6/6/14

40

Name _____   Date _____

Count the circles, and box the correct number. Color in the same number of circles on the right as the shaded ones on the left to show hidden partners.

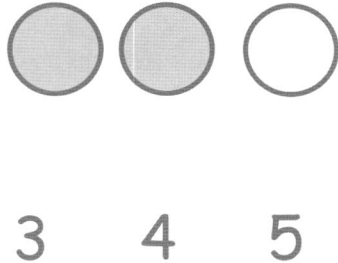

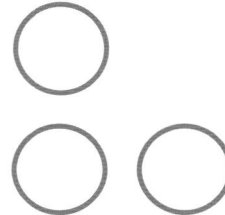

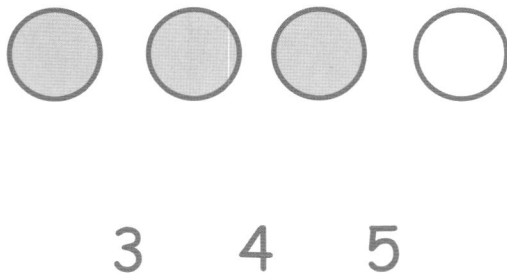

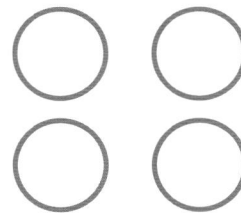

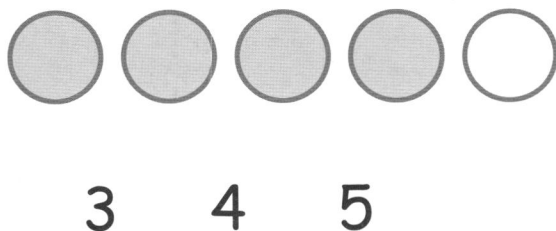

3   4   5

3   4   5

3   4   5

3   4   5

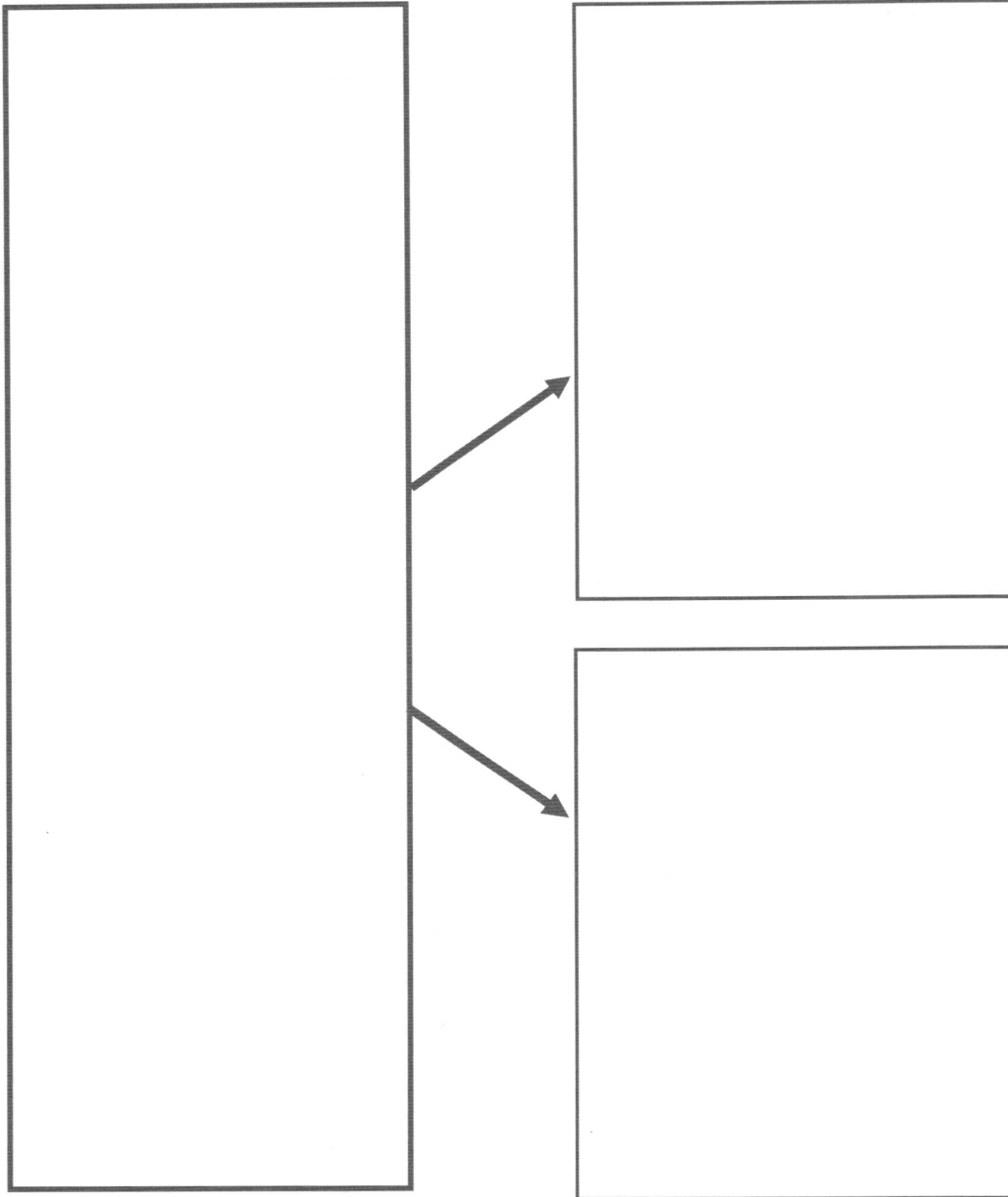

hidden partners

Lesson 9:   Within linear and array dot configurations of numbers 3, 4, and 5,
             find *hidden partners*.
Date: 6/6/14

42

Name _____   Date _____

Count the objects. Circle the total number of objects.
Color 1, 2, or 3 to see the hidden partners.

Color 1 circle.

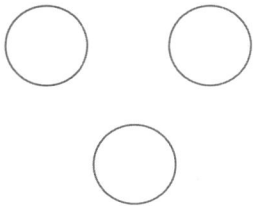

1   2   3

Color 3 stars.

2   3   4

Color 2 circles.

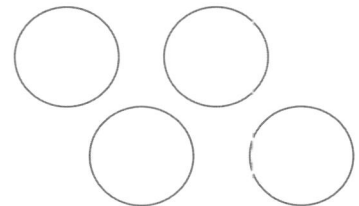

3   4   5

Color 3 circles.

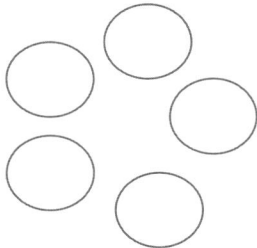

5   4   3

Color 4 stars.

4   5   3

Draw 2 circles and color them. Count all the objects, and circle the number.

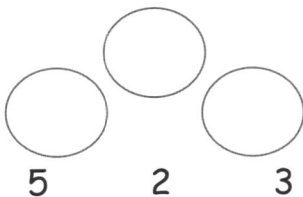

5   2   3

EUREKA MATH™   Lesson 10:   Within circular and scattered dot configurations of numbers 3, 4,
                              and 5, find *hidden partners*.   43
                 Date: 6/6/14

Name _____   Date _____

Count how many. Draw a box around that number. Then, circle a group of 3 dots in each box.

3 4 5	3 4 5
3 4 5	3 4 5
3 4 5	3 4 5

EUREKA MATH™   **Lesson 10:**   Within circular and scattered dot configurations of numbers 3, 4,   44
               and 5, find *hidden partners*.
               Date: 6/6/14

Name _____   Date _____

Count how many. Draw a box around that number. Then, color 3 of the circles in each group.

3   4   5   3   4   5

3   4   5   3   4   5

3   4   5   3   4   5

Talk to an adult at home about the hidden partners you found.

EUREKA MATH™   **Lesson 10:**   Within circular and scattered dot configurations of numbers 3, 4,   45
                          and 5, find *hidden partners*.
                Date: 6/6/14

large 5-frame cards

EUREKA MATH

Lesson 10:   Within circular and scattered dot configurations of numbers 3, 4, and 5, find *hidden partners*.

Date: 6/6/14

46

Name _____   Date _____

These squares represent cubes. Count the squares. Draw a line to break the stick between the gray squares and the white squares. Draw the squares above the numbers.

2   +   1

1   +   2

3   +   1

1   +   3

4   +   1

1   +   4

EUREKA MATH™

Lesson 11:   Model decompositions of 3 with materials, drawings, and expressions.
             Represent the decomposition as 1 + 2 and 2 + 1.
Date: 6/6/14

47

Name _____   Date _____

There are 2 green blocks and 1 yellow block. Draw the blocks.

```



```

There are 2 + 1 blocks. Count the blocks.

EUREKA
MATH™

Lesson 11:   Model decompositions of 3 with materials, drawings, and expressions.
               Represent the decomposition as 1 + 2 and 2 + 1.
Date: 6/6/14

48

Name _____   Date _____

Feed the puppies! Here are 3 bones. Draw lines to connect each bone with a puppy so that one puppy gets 2 bones and the other puppy gets 1 bone.

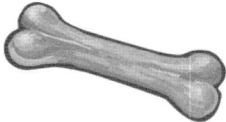

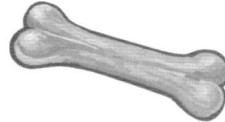

Color the shapes to show 1 + 4. Use your 2 favorite colors.

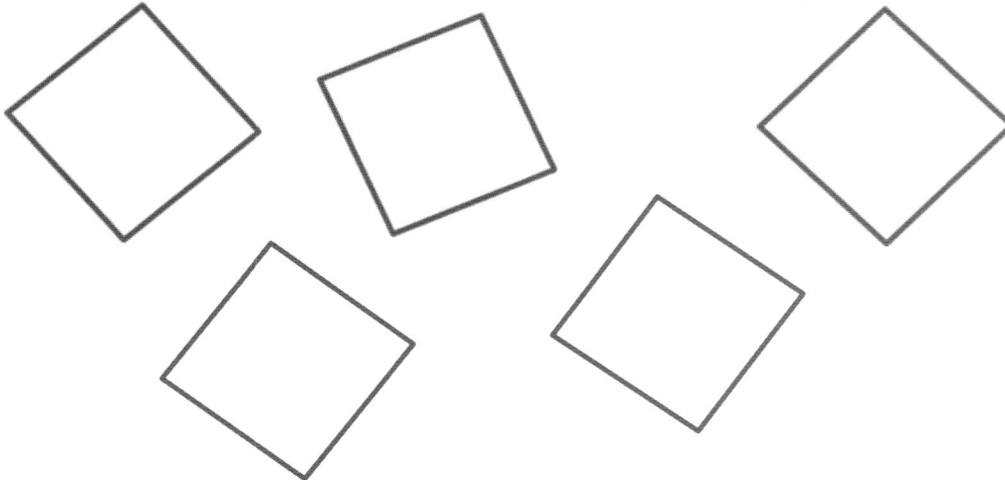

How many shapes are there? Circle the number.   1   2   3   4   5

EUREKA MATH   Lesson 11:   Model decompositions of 3 with materials, drawings, and expressions.   49
                                Represent the decomposition as 1 + 2 and 2 + 1.
                     Date: 6/6/14

Name _____   Date _____

Circle the number that tells how many.

0 1 2 3	0 1 2 3	0 1 2 3	0 1 2 3

Count the apples in each tree. Circle the number.

0 1 2 3	0 1 2 3	0 1 2 3	0 1 2 3

How many elephants are in the room?   _____

EUREKA MATH™   **Lesson 12:**   Understand the meaning of zero. Write the numeral 0.

Date: 6/6/14

50

Name _____   Date_____

✎Color in the blocks to show how many girls, boys, and aliens are at your table. Don't forget to count yourself!

Name _____　　Date_____

How many? Draw a line between each picture and its number.

| 0 |
| 1 |
| 2 |
| 3 |

Write the numbers in the blanks.

___ , 1, 2, 3

0, ___ , 2, 3

Name _____   Date _____

Insert the template into your personal white board. Practice with your dry erase marker. When you are ready, write in pencil on the paper.

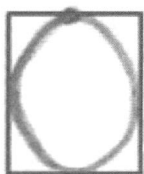     _____ _____

  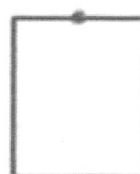   _____ _____

numeral formation practice sheet 0

Name _____   Date _____

Write the missing numbers.

1 2 ___	3 2 ___
1 ___ 3	3 ___ 1
0 1 ___	___ 2 1
0 ___ 2	2 1 ___
___ 1 2	2 ___ 0

EUREKA MATH   Lesson 13:   Order and write numerals 0–3 to answer *how many* questions.
Date: 6/6/14

54

Count and write how many.

_____      _____

_____   _____

Name _____   Date _____

Count the objects.

Write how many.

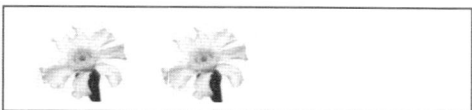

Fill in the missing numbers.

1, ____, 3   ____, 1, 2

3, 2, ____   ____, 1, 0

Name _____   Date _____

Draw ⬛⬛⬜⬜⬜ (two) things you see in your kitchen.

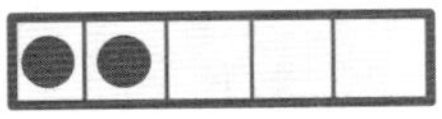

How many?

_____

Draw ⬛⬜⬜⬜⬜ (one) of your friends.

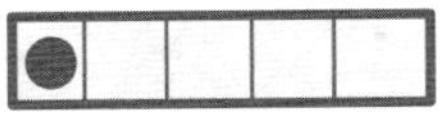

How many?

_____

Draw ⬛⬛⬛⬜⬜ (three) things you like to play.

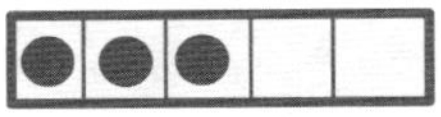

How many?

_____

How many pet monkeys 🐵 do you have? _____

Write the missing numbers:

3, 2, _____, _____   0, _____, _____, 3

EUREKA
MATH

Lesson 13:   Order and write numerals 0–3 to answer *how many* questions.
Date: 6/6/14

58

Name _____   Date _____

Insert this page into your personal white boards. Practice with your dry erase marker. When you are ready, write your numbers in pencil on the paper.

 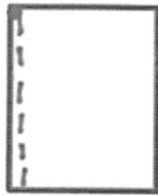  _____ _____

 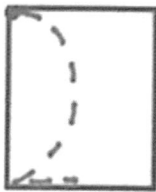  _____ _____

   _____ _____

_____

numeral formation practice sheet 1–3

Name _____ Date _____

Color the picture to match the number sentence.

3 = 1 + 2
Write the number sentence:

_____

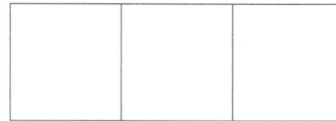

3 = 2 + 1
Write the number sentence:

_____

3 = 1 + 2
Write the number sentence:

_____

3 = 2 + 1
Write the number sentence:

_____

Look at the pictures above and write how many.

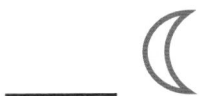 ____ 🌳 ____ ☆ ____ 🍎

EUREKA MATH™

Lesson 14: Write numerals 1–3. Represent decompositions with materials, drawings, and equations, 3 = 2 + 1 and 3 = 1 + 2.
Date: 6/6/14

60

Name _____   Date _____

Color the apples to show that 3 = 2 + 1.

How many apples are there altogether? _____

3 is the same as _____ and _____.

3 apples = _____ apples + _____ apple

EUREKA MATH™

Lesson 14:   Write numerals 1–3. Represent decompositions with materials,
              drawings, and equations, 3 = 2 + 1 and 3 = 1 + 2.
Date: 6/6/14

61

Name _____   Date _____

Color the shirts so that 1 is red and 2 are green. There are _____
shirts. _____ = 1 + _____

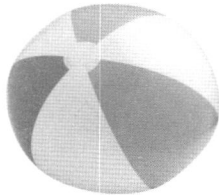

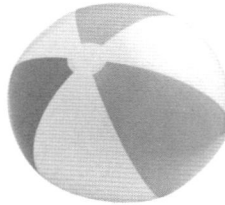

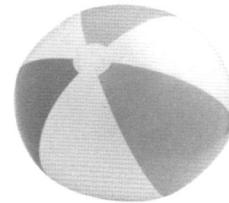

Color the balls so that 2 are yellow and 1 is blue. There are _____
balls. _____ = 2 + _____

Choose two of your favorite types of fruit. Draw some of each on the
plate to show that 3 = 2 + 1.

_____ fruits = _____ fruits + _____ fruit

_____ = _____ + _____

EUREKA
MATH

Lesson 14:   Write numerals 1–3. Represent decompositions with materials,
             drawings, and equations, 3 = 2 + 1 and 3 = 1 + 2.
Date: 6/6/14

62

Name _____   Date _____

Count and write how many. Circle a group of four of each fruit.

EUREKA MATH

Lesson 15:   Order and write numerals 4 and 5 to answer *how many* questions in
               categories; sort by count.
Date: 6/6/14

Name _____ Date _____

How many?

_____

How many?

_____

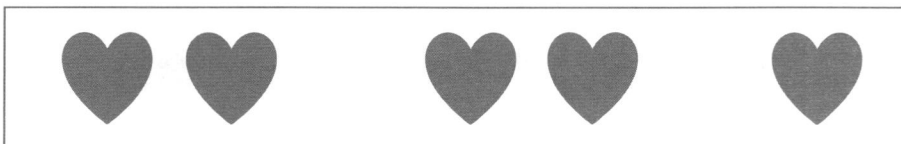

Are there more ♥ or ★ ? Circle the shape that has more.

Write the missing numbers:

1, 2, 3, _____, _____

EUREKA MATH

Lesson 15: Order and write numerals 4 and 5 to answer *how many* questions in categories; sort by count.
Date: 6/6/14

64

Name _____   Date _____

Count the shapes and write the numbers. Mark each shape as you count.

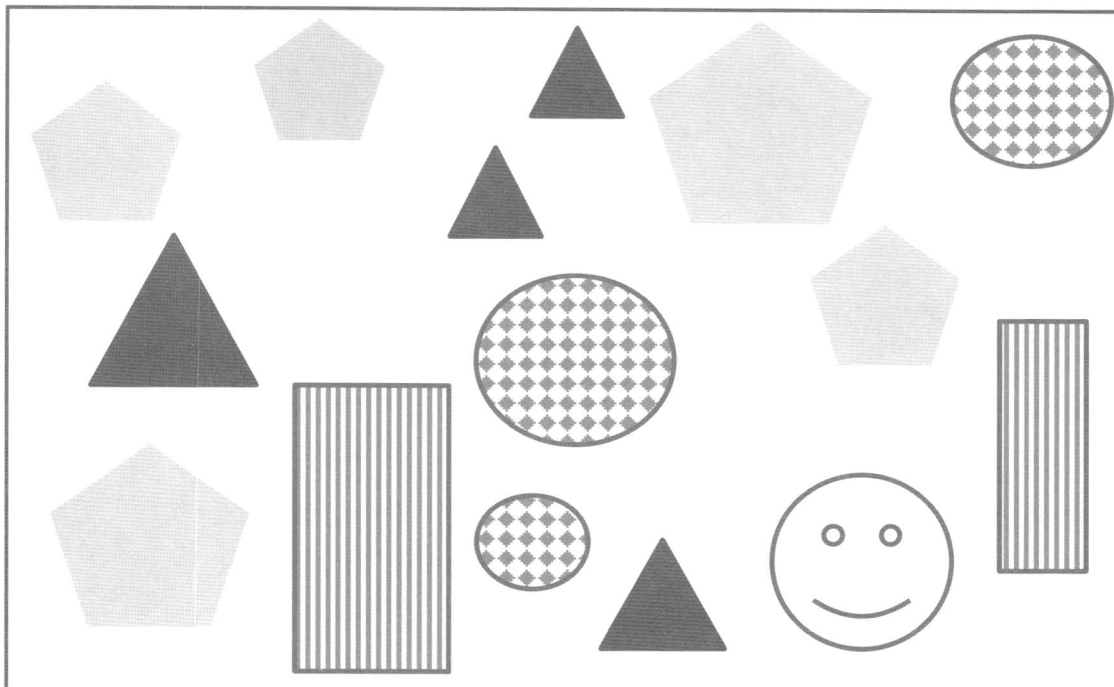

How many?   _____   _____   _____

              _____   _____   _____

Write the missing numbers:

0, 1, _____, 3, _____, _____

_____, _____, 3, 2, 1, _____

| 1 | 2 | 3 | 4 | 5 | 6 | 7 | 8 | 9 | 10 |

| 1 | 2 | 3 | 4 | 5 | 6 | 7 | 8 | 9 | 10 |

| 1 | 2 | 3 | 4 | 5 | 6 | 7 | 8 | 9 | 10 |

| 1 | 2 | 3 | 4 | 5 | 6 | 7 | 8 | 9 | 10 |

| 1 | 2 | 3 | 4 | 5 | 6 | 7 | 8 | 9 | 10 |

number path

EUREKA MATH™

Lesson 15:   Order and write numerals 4 and 5 to answer *how many* questions in categories; sort by count.
Date: 6/6/14

66

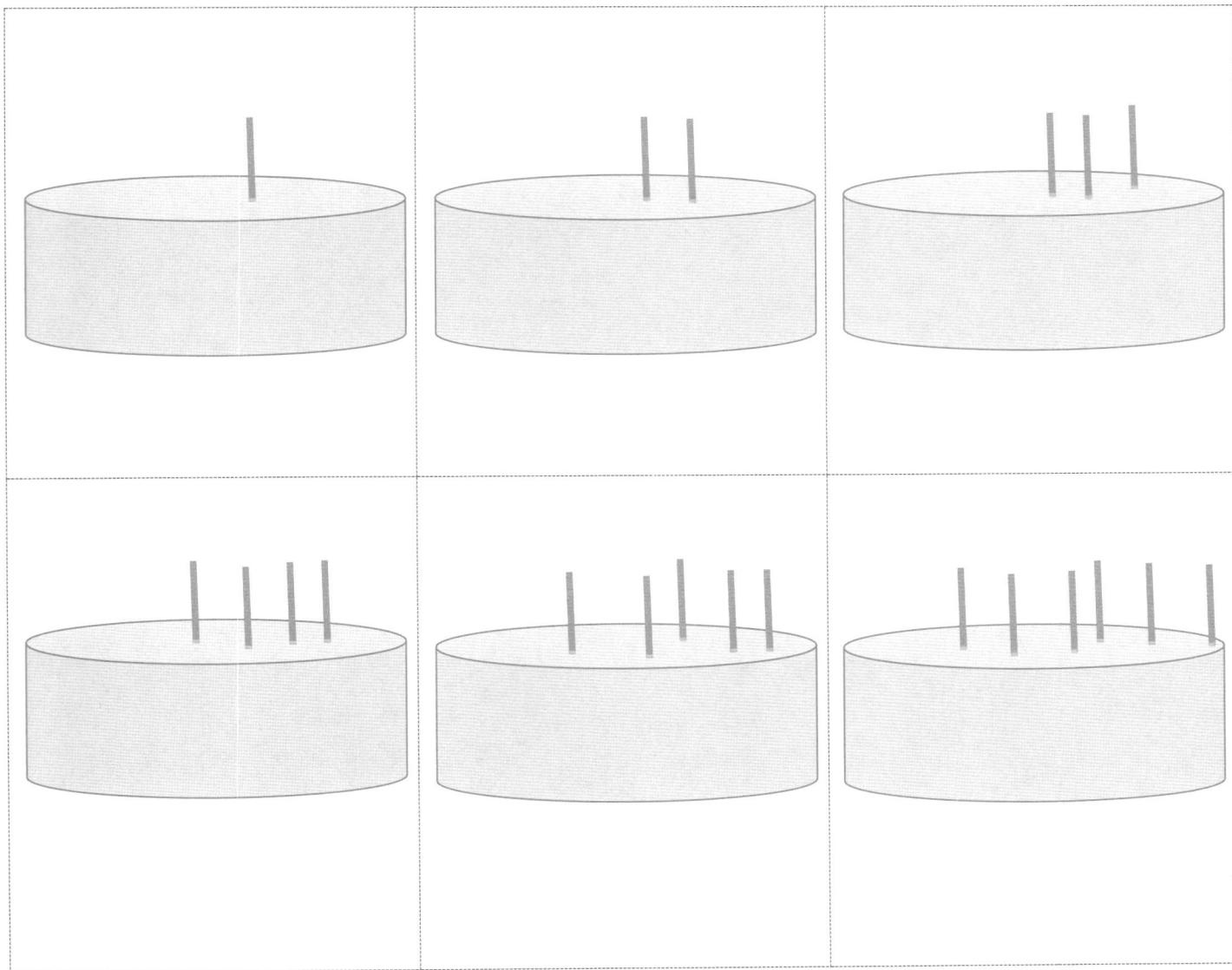

birthday cake number order cards

**EUREKA
MATH™**

Lesson 15:   Order and write numerals 4 and 5 to answer *how many* questions in
categories; sort by count.
Date: 6/6/14

67

Name _____   Date _____

Insert the template into your personal white board. Practice with your dry
erase marker. When you are ready, write in pencil on the paper.

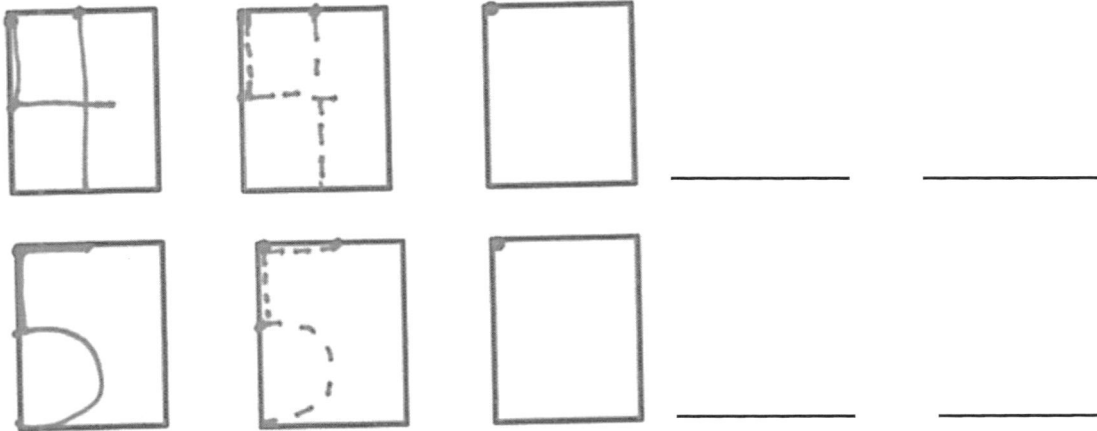

_____   _____

_____   _____

Write the missing numbers:

_____, 2, 3, _____, _____

_____, 4, 3, _____, _____

1, 2, _____, _____, _____

_____
numeral formation practice sheet 4–5

EUREKA
MATH

Lesson 15:   Order and write numerals 4 and 5 to answer *how many* questions in
              categories; sort by count.
Date: 6/6/14

68

Name _____   Date _____

In each picture, color some squares red and some blue. Do it a different way each time.

How many squares? _____

How many squares? _____

How many squares? _____

How many squares? _____

Draw more circles to make 4.

OOO	OO	O

Draw more X's to make 5.

XXXX	XXX	XX	X

EUREKA
MATH

Lesson 16:   Write numerals 1–5 in order. Answer and make drawings of
decompositions with totals of 4 and 5 without equations.
Date: 6/6/14

69

Name _____   Date _____

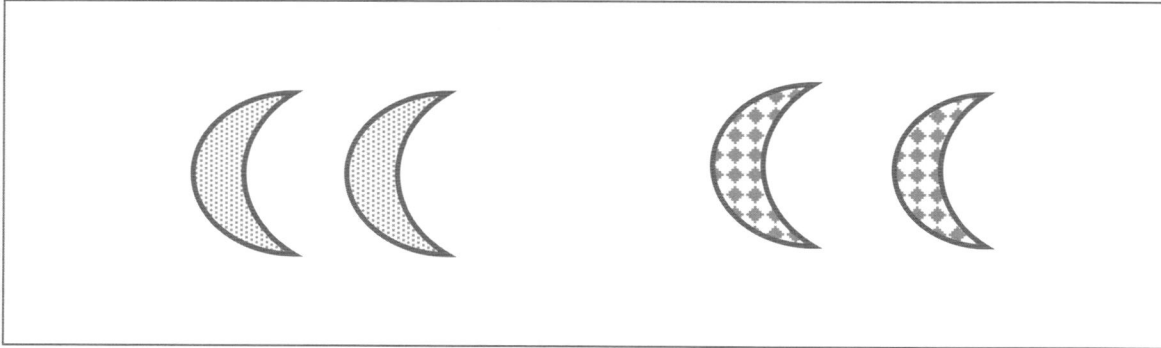

How many 🌙 ? _____   How many 🌙 ? _____

How many altogether? _____

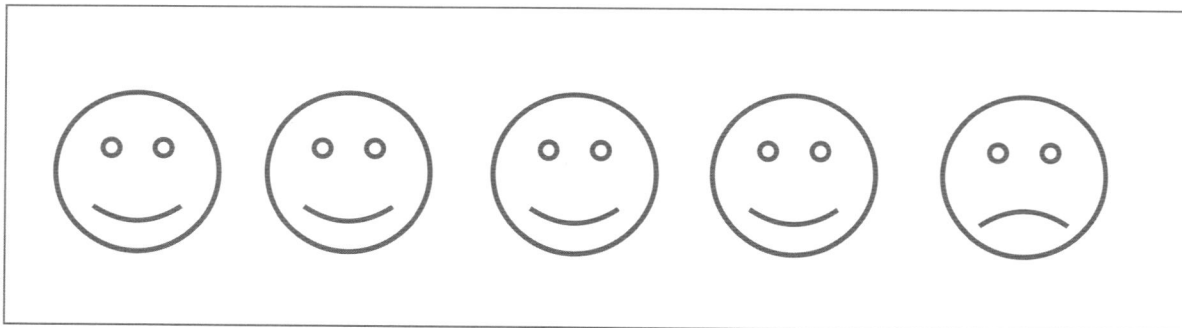

How many 🙂 ? _____   How many 🙁 ? _____

How many altogether? _____

Name _____   Date _____

Write the missing numbers:

1, 2, ___, 4, ___   5, ___, 3, 2, ___

___, 3, 2, 1, ___   ___, 1, 2, ___, 4

Draw 3 red fish and 1 green fish.

How many fish are there in all? There are _____ fish.

3 fish and 1 fish make _____ fish. 4 is the same as _____ and _____.

Make 2 happy faces and 3 sad faces.

How many faces are there in all? There are _____ faces.

2 faces and 3 faces make _____ faces.

5 is the same as _____ and _____.

EUREKA MATH™   Lesson 16:   Write numerals 1–5 in order. Answer and make drawings of
                    decompositions with totals of 4 and 5 without equations.   71
            Date: 6/6/14

© 2014 Common Core, Inc. All rights reserved. commoncore.org

Name _____   Date _____

Draw 1 more. Then, count the objects and write the number in the box.
Use the code to color when you are finished.

| 3 blue | 4 red | 5 yellow | 6 green |

Draw 1 more cloud.

How many? [ ]

Draw 1 more face.

How many? [ ]

Draw 1 more heart.

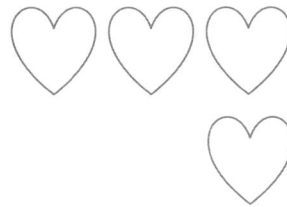

How many? [ ]

Draw 1 more.
Then, circle the number.

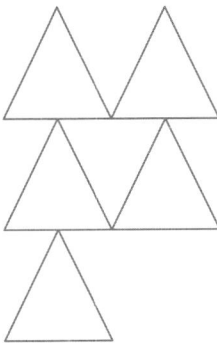

4   5   6

Draw 6 fingers.

Draw 6 beads.

EUREKA MATH™   **Lesson 17:** Count 4–6 objects in vertical and horizontal linear configurations and array configurations. Match 6 objects to the numeral 6.
Date: 6/6/14

Name _____   Date _____

Fill in the missing numbers on the cards.

| 0 | 1 | | 3 | | 5 | 6 |

Count. Write how many in the box.

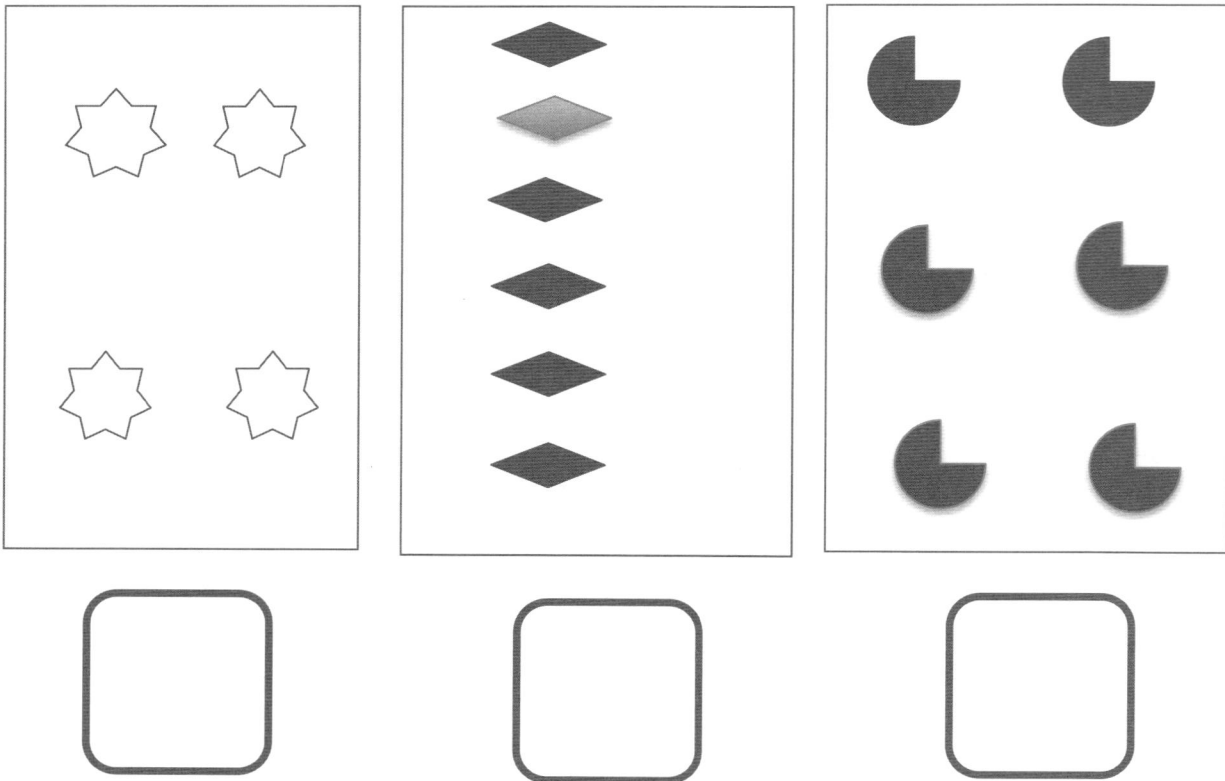

EUREKA
MATH™

Lesson 17:   Count 4–6 objects in vertical and horizontal linear configurations and
array configurations. Match 6 objects to the numeral 6.
Date: 6/6/14

Name _____ Date _____

Color 4.

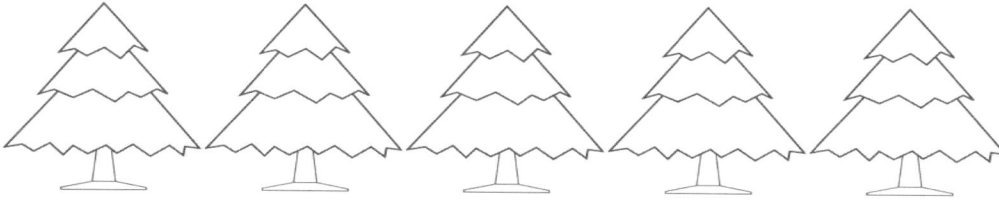

Color 5.

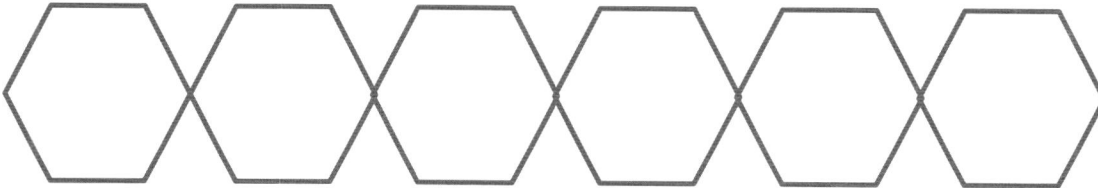

Color 6.

Connect the boxes with the same number.

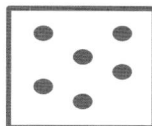

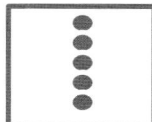

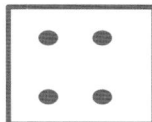

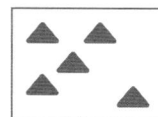

EUREKA MATH

Lesson 17: Count 4–6 objects in vertical and horizontal linear configurations and
array configurations. Match 6 objects to the numeral 6.
Date: 6/6/14

74

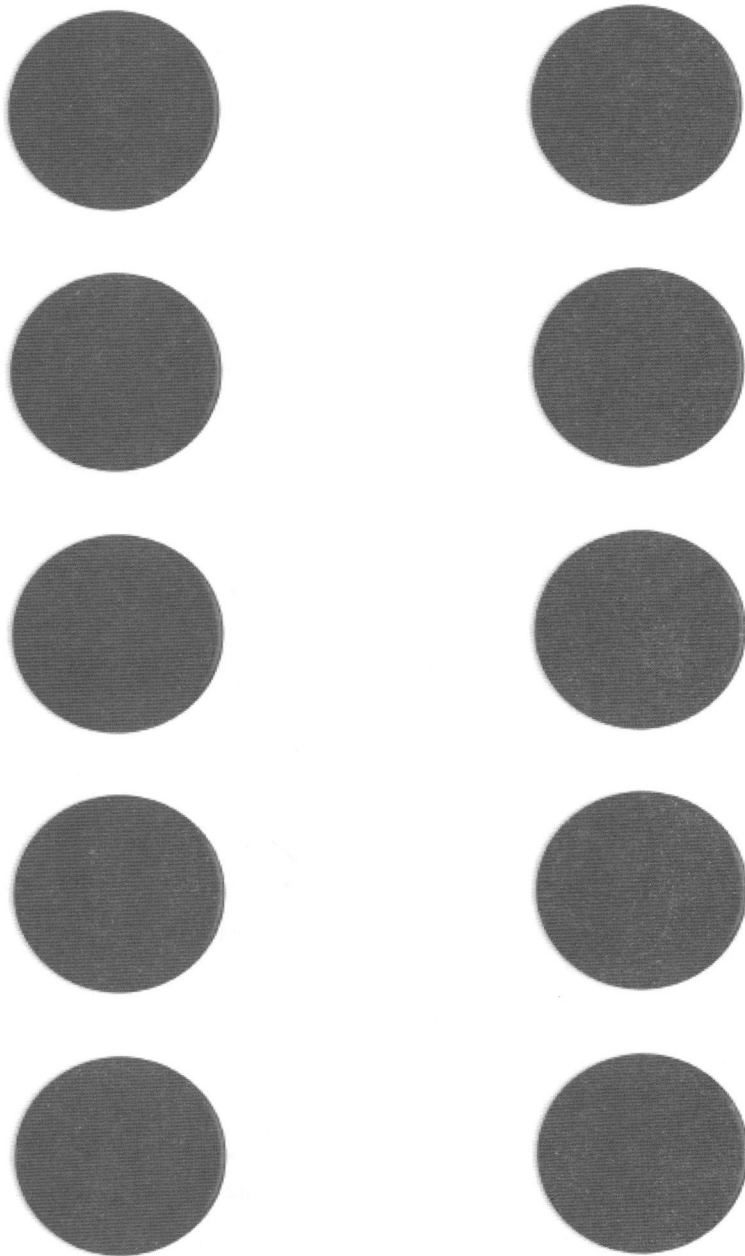

5-group mat

EUREKA
MATH

Lesson 17:   Count 4–6 objects in vertical and horizontal linear configurations and
                 array configurations. Match 6 objects to the numeral 6.
Date: 6/6/14

75

Name_____Date_____

Insert this page into your personal whiteboards.   Practice. When you are ready, write your numbers in pencil on the paper.

   _____ _____

   _____ _____

EUREKA MATH™

Lesson 18:   Count 4–6 objects in circular and scattered configurations.
Count 6 items out of a larger set. Write numerals 1–6 in order.
Date: 6/6/14

76

Name _____   Date _____

Color 6 beans.   Color 6 beans.

   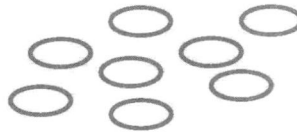

Color 6 beans.   Color 6 beans.

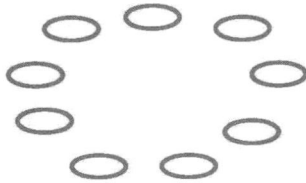   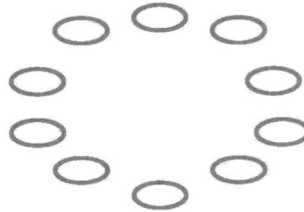

Count the dots in each box. Write the number in the box.

EUREKA MATH™ | **Lesson 18:** | Count 4–6 objects in circular and scattered configurations.
                   | | Count 6 items out of a larger set. Write numerals 1–6 in order.
                   | **Date:** | 6/6/14

77

Count the objects. Write the number in the box.

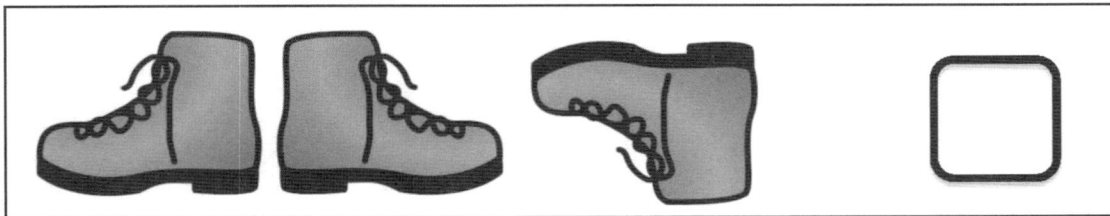

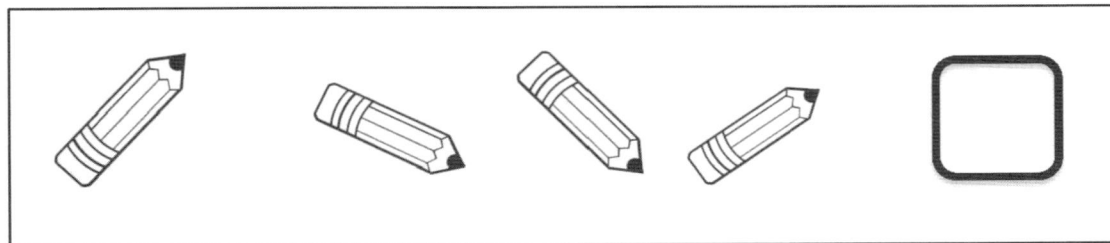

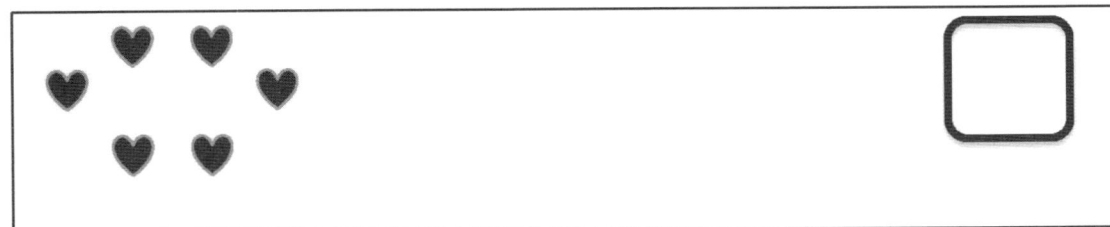

EUREKA
MATH

Lesson 18:   Count 4–6 objects in circular and scattered configurations.
             Count 6 items out of a larger set. Write numerals 1–6 in order.
Date: 6/6/14

78

Name _____ Date _____

Draw 6 beads on this magic necklace.

Fill in the missing numbers.

_____, 2, _____, _____, 5, _____

EUREKA MATH™

Lesson 18: Count 4–6 objects in circular and scattered configurations.
Count 6 items out of a larger set. Write numerals 1–6 in order.
Date: 6/6/14

79

Name _____   Date _____

Color 6   Color 5

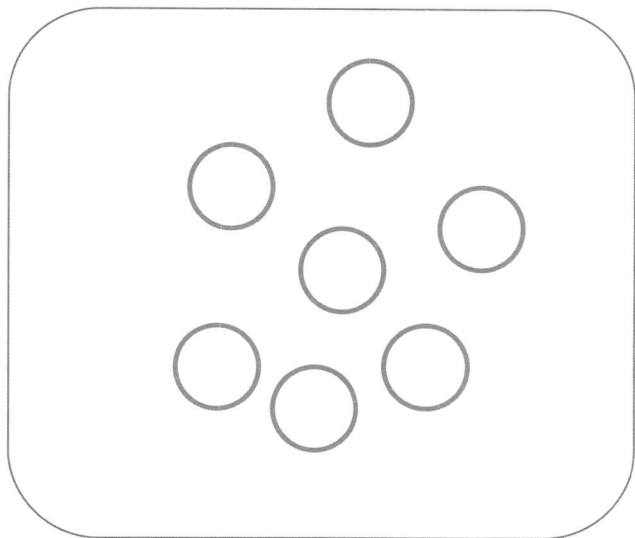   

Circle 6  balloons.

EUREKA MATH   **Lesson 18:** Count 4–6 objects in circular and scattered configurations.   80
                                Count 6 items out of a larger set. Write numerals 1–6 in order.
                Date: 6/6/14

Name _____   Date _____

Color 5

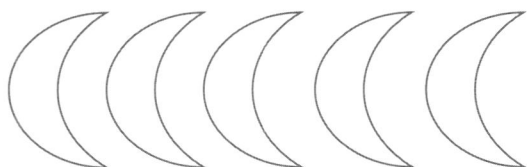

Color 5

Color 5

Color 5

Color 5. Draw 2 circles to the right. Write the total.

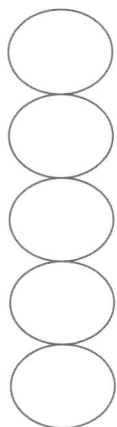

Color 5. Draw 2 circles below. Write the total.

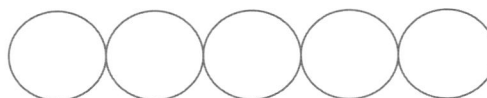

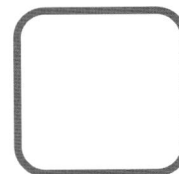

EUREKA MATH™

Lesson 19:   Count 5–7 linking cubes in linear configurations. Match with numeral 7. Count on fingers from 1 to 7 and connect to 5-group images.
Date: 6/6/14

81

Name _____   Date _____

Color 5 squares on the 5-group card. Then, color 2 squares on the other 5-group card.

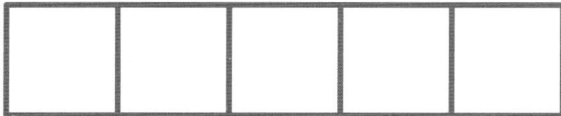

Count how many squares you colored.

Write the numeral in the box.

Answer my riddle. 7 is 5 and _____ more.

EUREKA MATH

Lesson 19: Count 5–7 linking cubes in linear configurations. Match with numeral 7. Count on fingers from 1 to 7 and connect to 5-group images.

Date: 6/6/14

82

Name _____   Date _____

Draw a line from the numeral to the 5-group cards it matches.

3

4

5

6

7

Fill in the missing numbers.

___, 5, ___, 7

7, 6, ___, 4, ___, 2

1, ___, 3, ___, 5, ___, ___

7, ___, 5, ___, ___, 2, 1

EUREKA MATH™

Lesson 19:   Count 5–7 linking cubes in linear configurations. Match with numeral
             7. Count on fingers from 1 to 7 and connect to 5-group images.
Date: 6/6/14

83

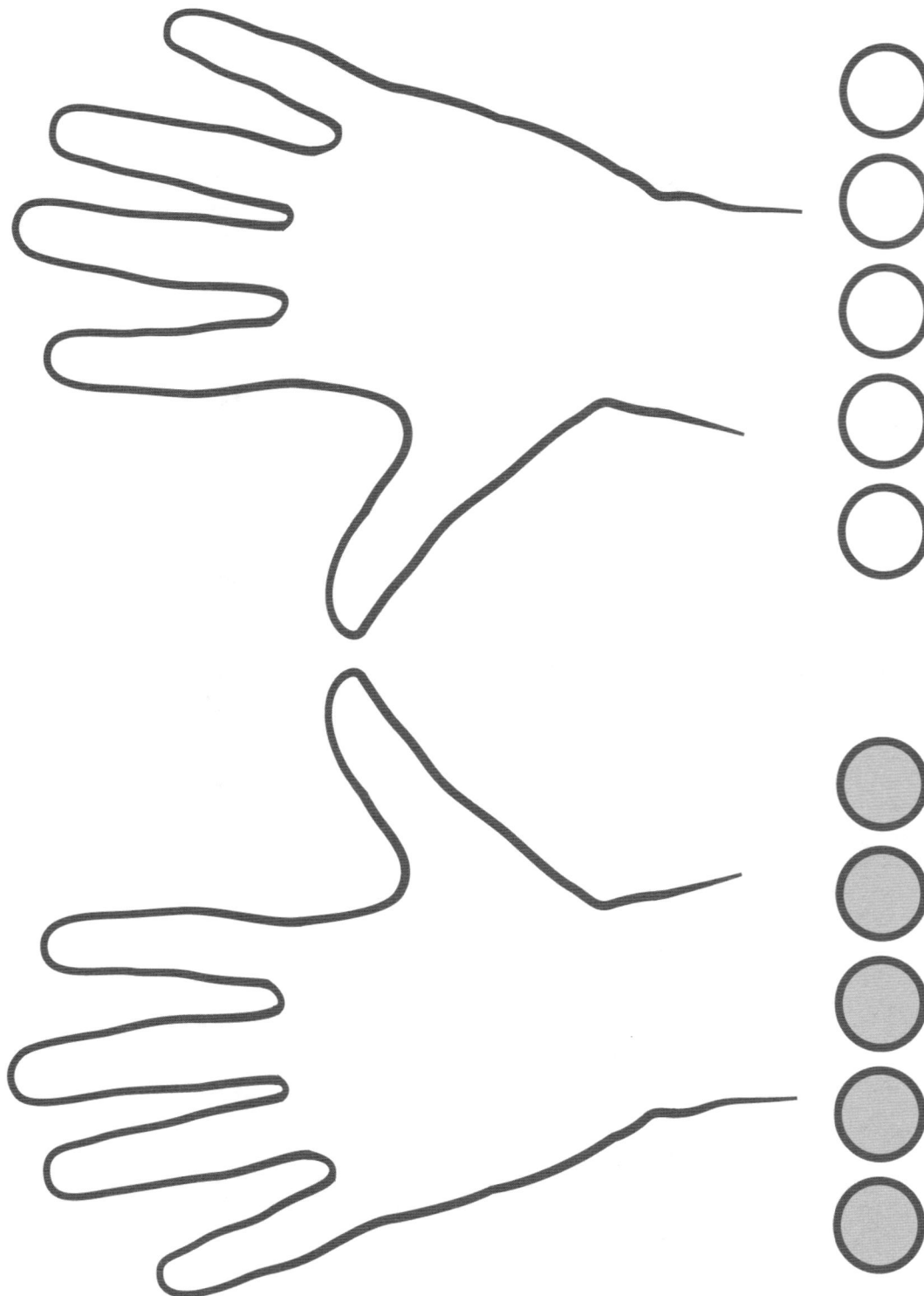

two hands mat

EUREKA MATH

Lesson 19:   Count 5–7 linking cubes in linear configurations. Match with numeral
             7. Count on fingers from 1 to 7 and connect to 5-group images.
Date: 6/6/14

84

Name_____Date_____

Insert this page into your personal whiteboards.   Practice. When you are ready, write your numbers in pencil on the paper.

  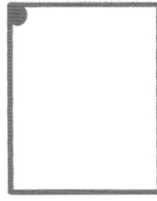   _____   _____

     _____   _____

EUREKA MATH

Lesson 20:   Reason about sets of 7 varied objects in circular and scattered configurations. Find a path through the scattered configuration. Write numeral 7. Ask, "How is your seven different from mine?"

Date: 6/6/14

85

Name _____   Date _____

Color 7 beans. Draw a line to connect the dots you colored.

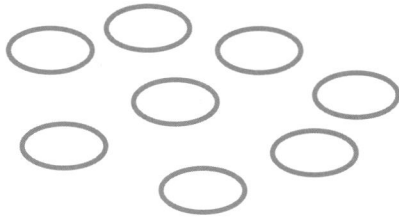   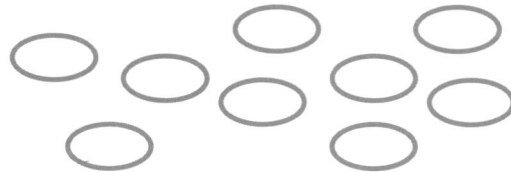

Color 7 beans.   Color 7 beans.

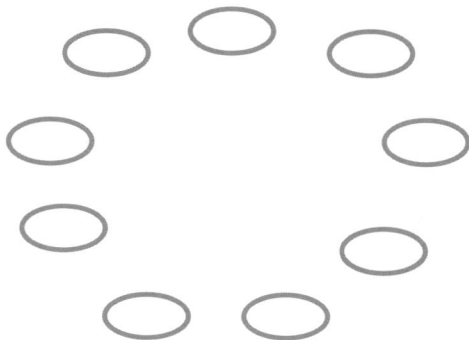   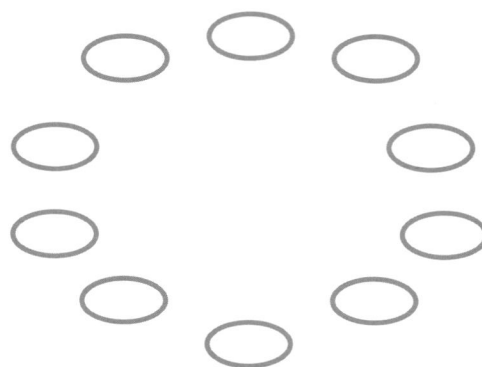

Count the dots in each box. Write the number in the box.

EUREKA MATH™

Lesson 20:   Reason about sets of 7 varied objects in circular and scattered configurations. Find a path through the scattered configuration. Write numeral 7. Ask, "How is your seven different from mine?"

Date: 6/6/14

86

Name _____   Date _____

Make a necklace. Draw 7 beads around the circle.

Color 7 ♡ hearts.

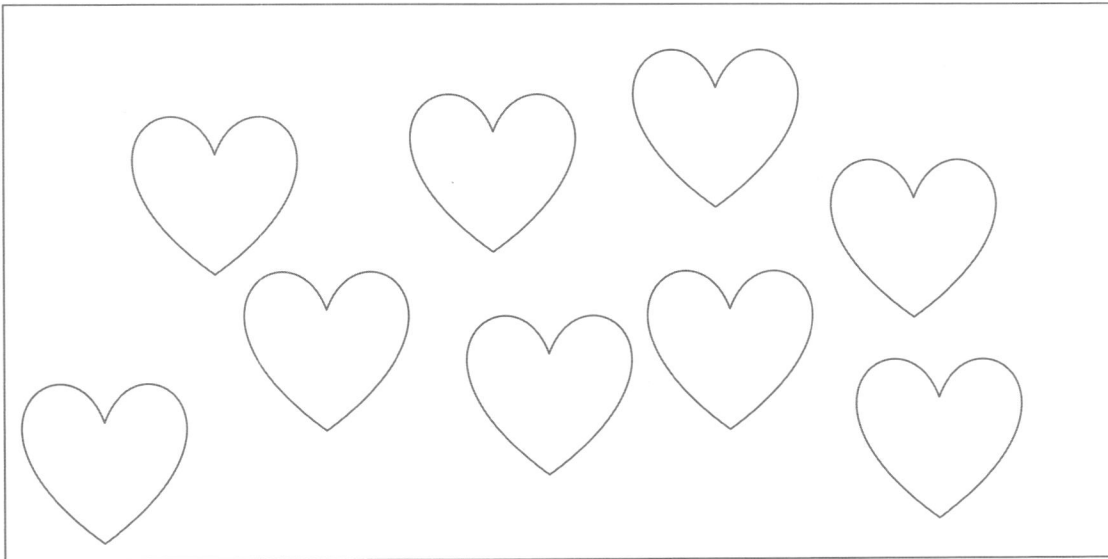

EUREKA
MATH

Lesson 20: Reason about sets of 7 varied objects in circular and scattered configurations. Find a path through the scattered configuration. Write numeral 7. Ask, "How is your seven different from mine?"

Date: 6/6/14

Name _____   Date _____

How many? Write the number in the box.

Lesson 20:   Reason about sets of 7 varied objects in circular and scattered
configurations. Find a path through the scattered configuration.
Write numeral 7. Ask, "How is your seven different from mine?"
Date: 6/6/14

88

Count how many. Write the number in the box.
Draw a line to show how you counted the suns.

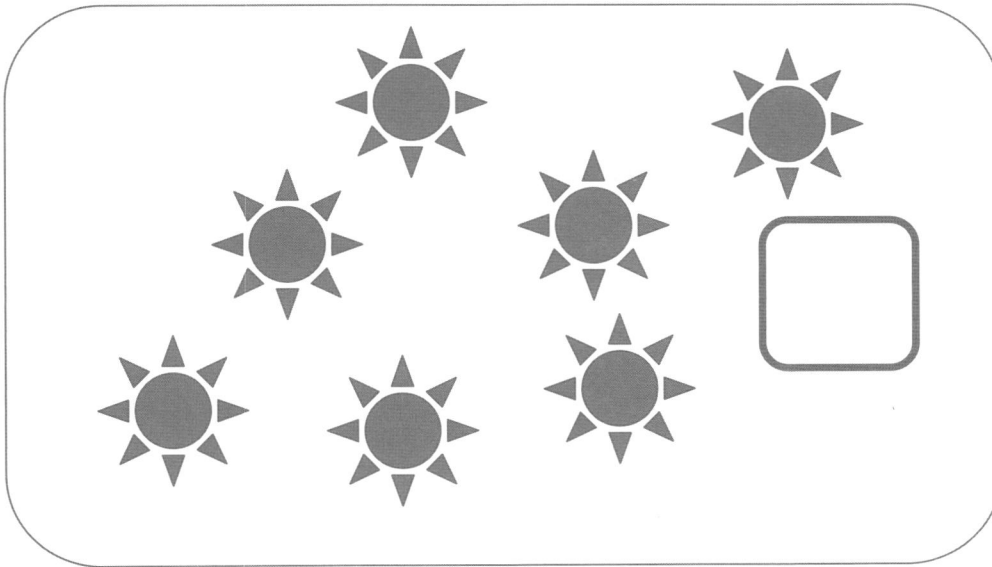

Count how many. Write the number in the box.
Draw a line to show how you counted the circles.

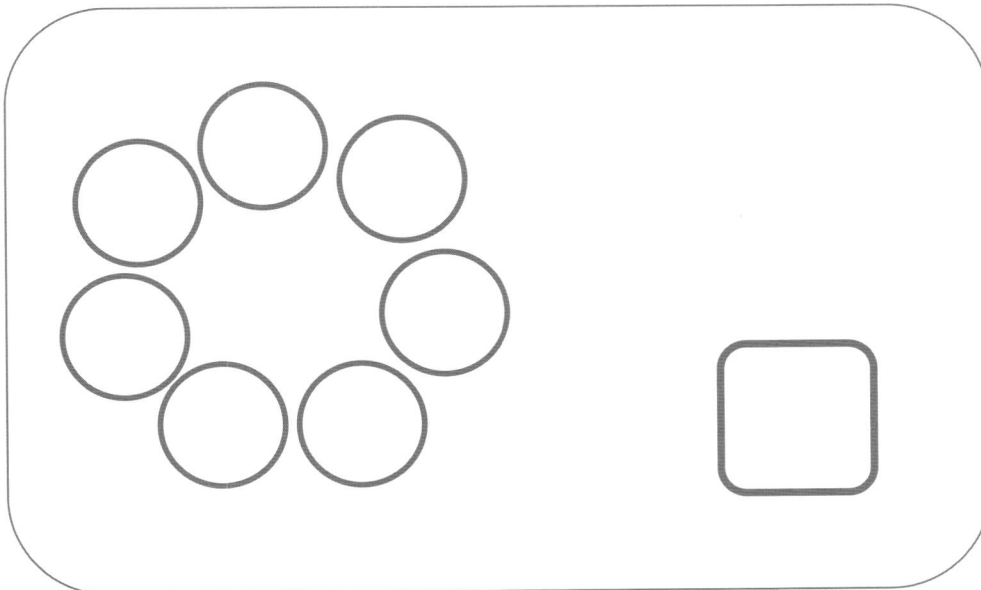

EUREKA MATH

Lesson 20: Reason about sets of 7 varied objects in circular and scattered
configurations. Find a path through the scattered configuration.
Write numeral 7. Ask, "How is your seven different from mine?"

Date: 6/6/14

Name _____   Date _____

Color 5 ladybugs. Color the remaining ladybugs a different color.
Count all the ladybugs, and write how many.

Color 5 diamonds. Color the remaining diamonds a different color.
Count all the diamonds, and write how many.

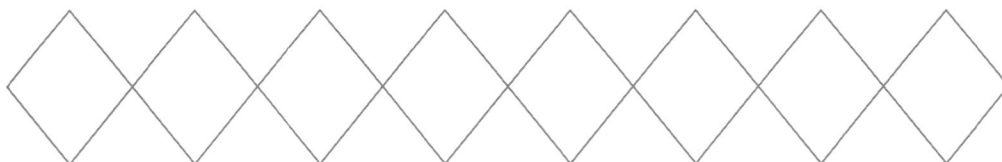

Color 5 circles. Then, draw 3 circles
to the right. Count all the circles.
Write how many in the box.

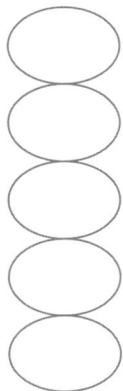

Color 5 circles. Then, draw 3 circles
below. Count all the circles. Write
how many in the box.

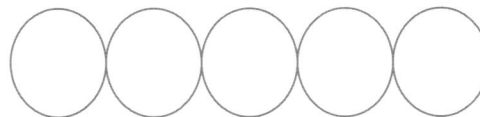

Color 4 ladybugs. Count all the ladybugs, and write how many in the box.

Color 4 rectangles. Count all the rectangles, and write how many in the box.

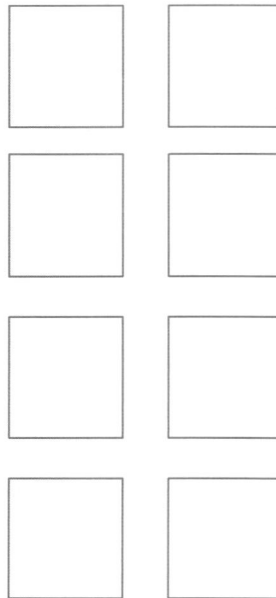

Color 5. Then, draw 3 circles to finish the row. Color the bottom 3 circles you drew a different color. Write the total in the box.

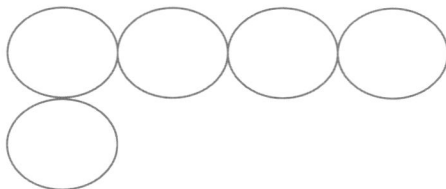

EUREKA MATH

Lesson 21:   Compare counts of 8 in linear and array configurations. Match with numeral 8.
Date: 6/6/14

91

© 2014 Common Core, Inc. All rights reserved. commoncore.org

Name _____   Date _____

Color 4 squares red and 4 squares blue. Count all the squares. Write how many in the box.

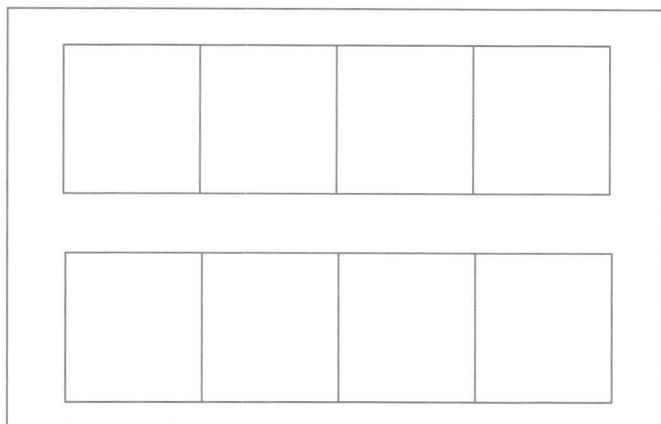

Color 6 squares red and 2 squares blue. Write the number of squares in the box.

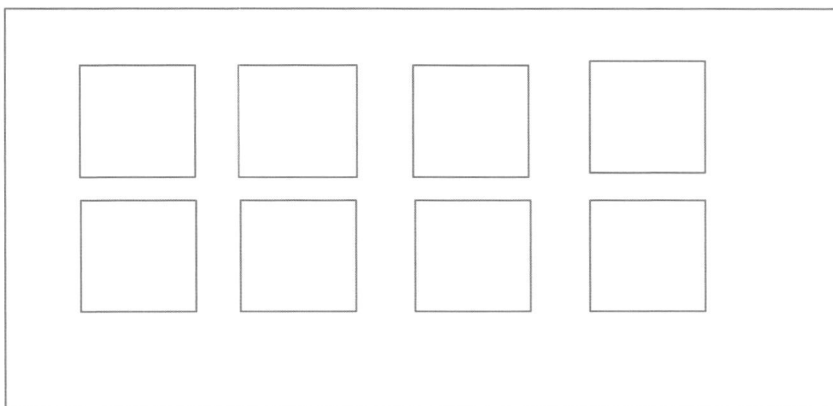

EUREKA MATH™

Lesson 21:   Compare counts of 8 in linear and array configurations. Match with numeral 8.
Date: 6/6/14

92

Name _____   Date _____

Color 4 squares blue. Color 4 squares yellow.

Count how many squares. Write the number in the box.

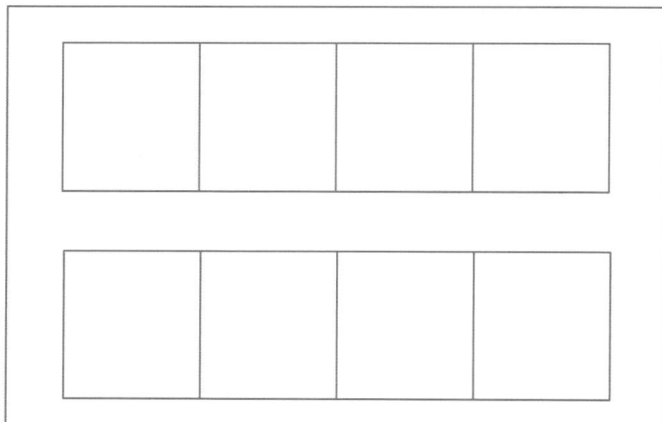

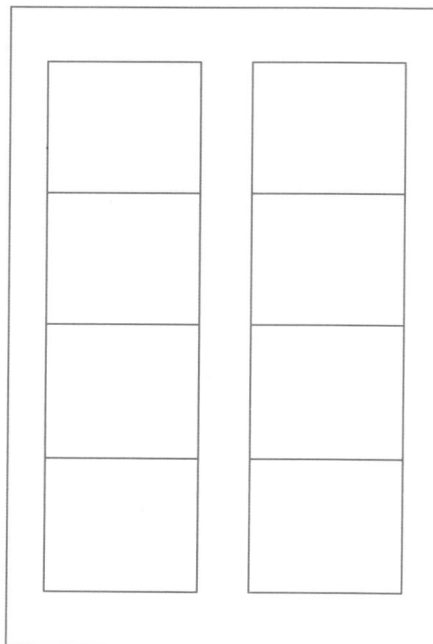

Color 4 squares green. Color 4 squares brown.

Count how many squares. Write the number in the box.

EUREKA MATH   **Lesson 21:**   Compare counts of 8 in linear and array configurations. Match with numeral 8.

Date: 6/6/14   93

Count how many. Write the number in the box.

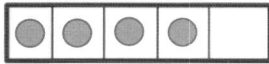

EUREKA
MATH™

Lesson 21: Compare counts of 8 in linear and array configurations. Match with numeral 8.
Date: 6/6/14

94

Name _____   Date _____

Insert this page into your personal white boards. Practice. When you are ready, write your numbers in pencil on the paper.

   _____ _____

 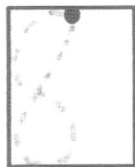  _____ _____

Color 8 happy faces.

Circle a different group of 8 happy faces.

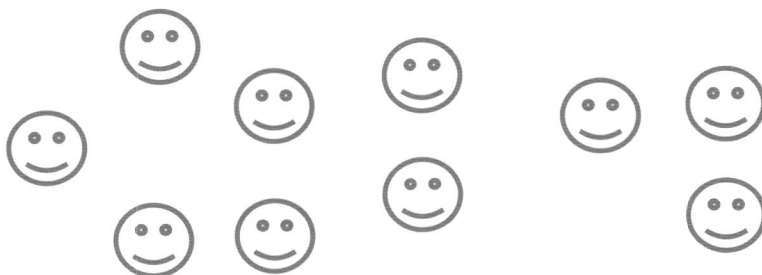

EUREKA MATH™ | **Lesson 22:**   Arrange and strategize to count 8 beans in circular (around a cup) and scattered configurations. Write numeral 8. Find a path through the scatter set and compare paths with a partner.   95

Date: 6/6/14

Name _____   Date _____

Draw a counting path with a line to show the order in which you counted.
Write the total number in the box. Circle a group of 5 in each set.

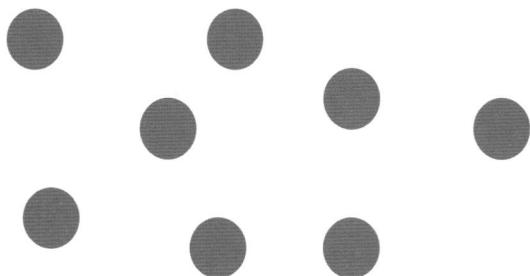

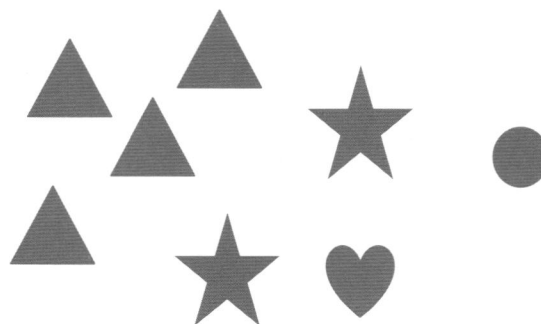

Number the circles from 1 to 8.
Color 8 circles.

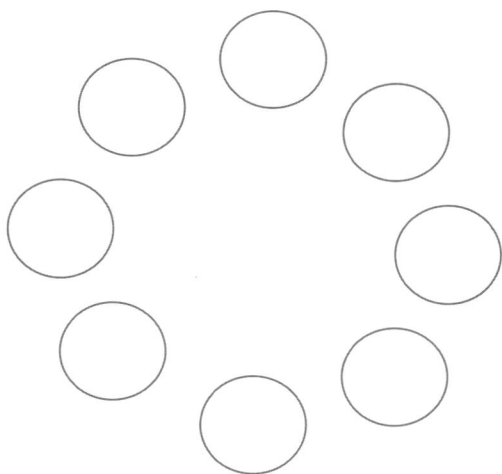

Number the shapes from 1 to 8.

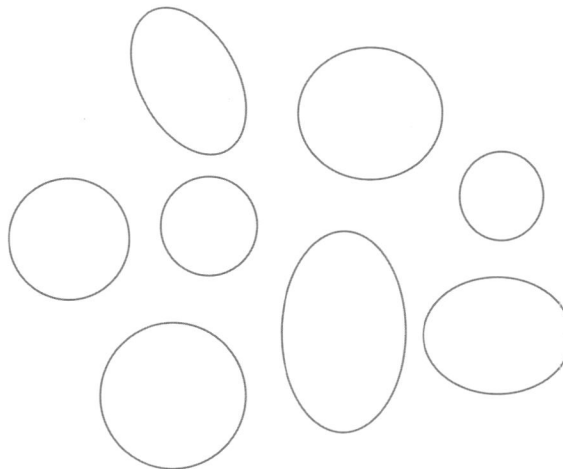

EUREKA MATH™   Lesson 22:   Arrange and strategize to count 8 beans in circular (around a cup) and
scattered configurations. Write numeral 8. Find a path through the
scatter set and compare paths with a partner.

Date: 6/6/14

96

Name _____   Date _____

Count. Write the number in the box.

Draw a line to show your counting path.

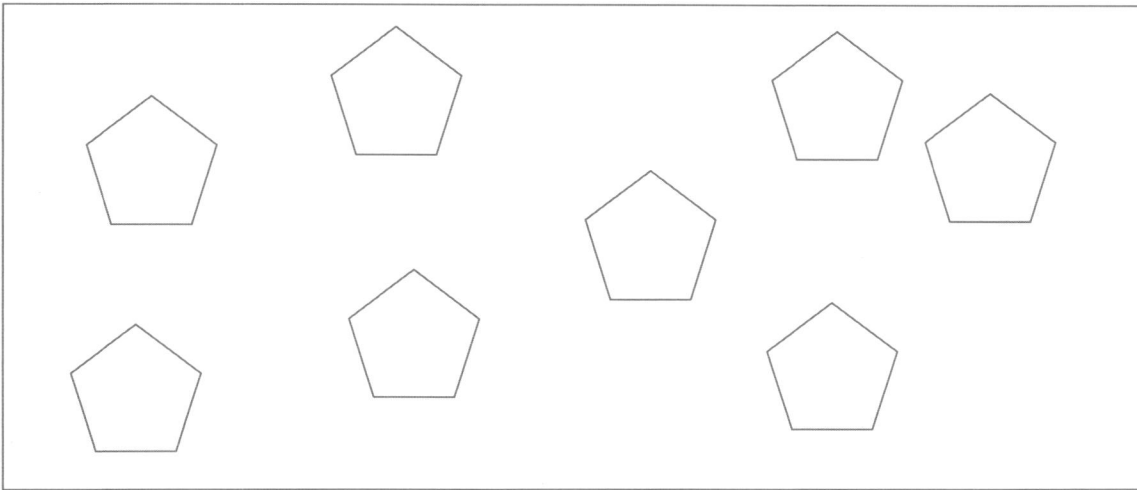

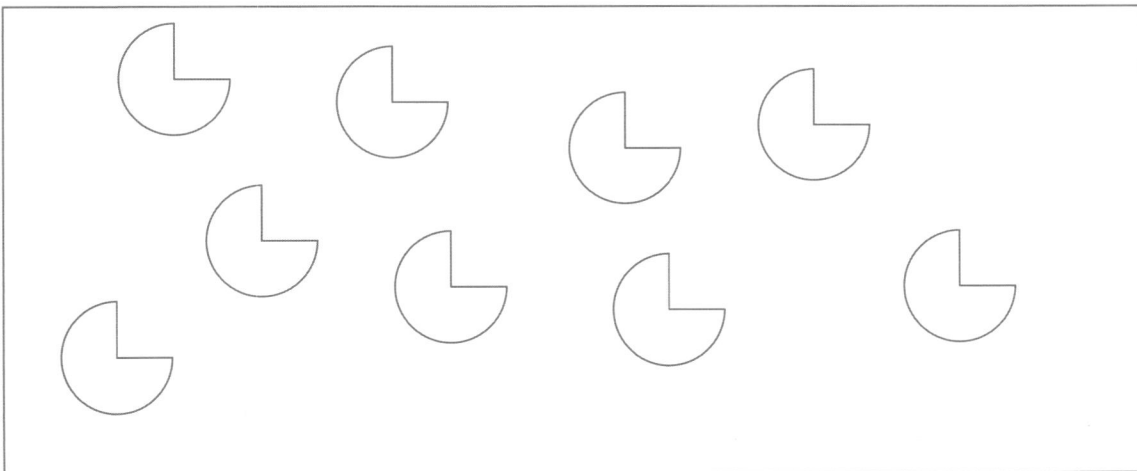

Color 8.

Draw a line to show your counting path.

EUREKA MATH

Lesson 22: Arrange and strategize to count 8 beans in circular (around a cup) and scattered configurations. Write numeral 8. Find a path through the scatter set and compare paths with a partner.

Date: 6/6/14

97

Name _____ Date _____

Draw 8 beads around the circle.

Color 8. Draw a line to show your counting path.

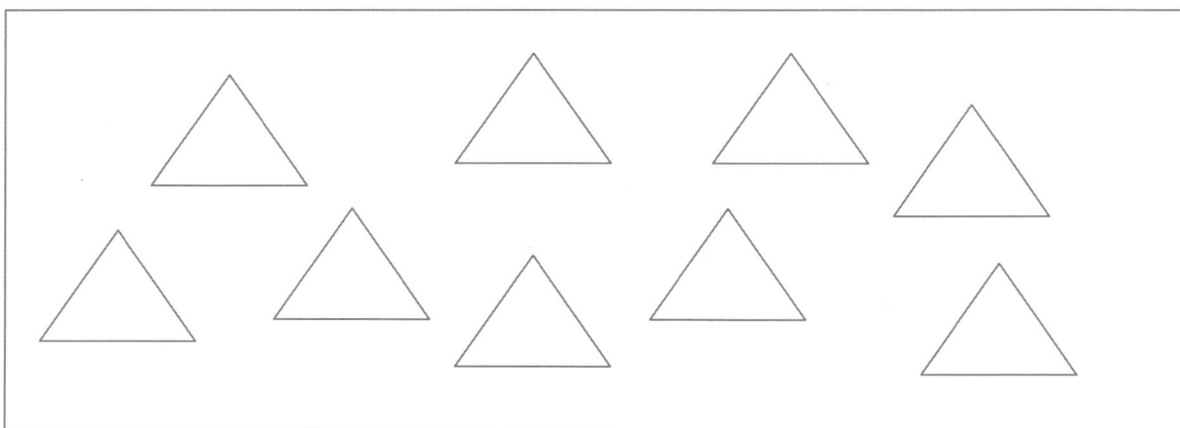

EUREKA
MATH

Lesson 22:   Arrange and strategize to count 8 beans in circular (around a cup) and
scattered configurations. Write numeral 8. Find a path through the
scatter set and compare paths with a partner.
Date: 6/6/14

98

Count how many. Write the number in the box.

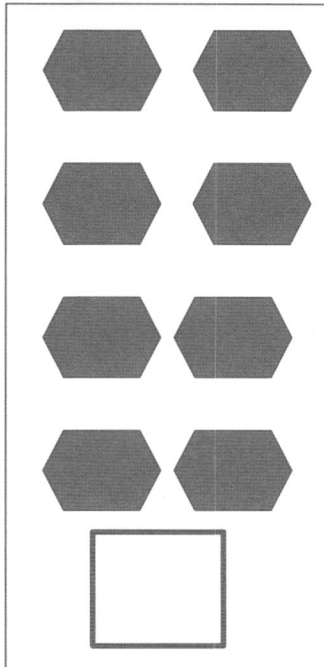

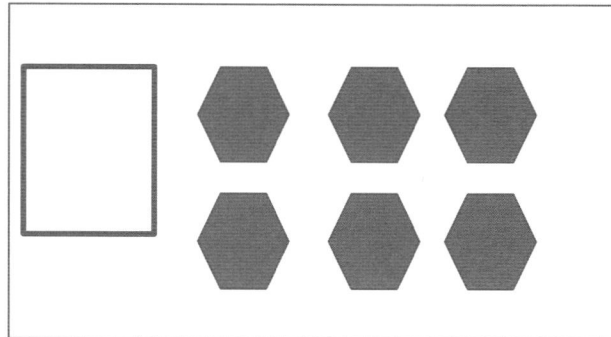

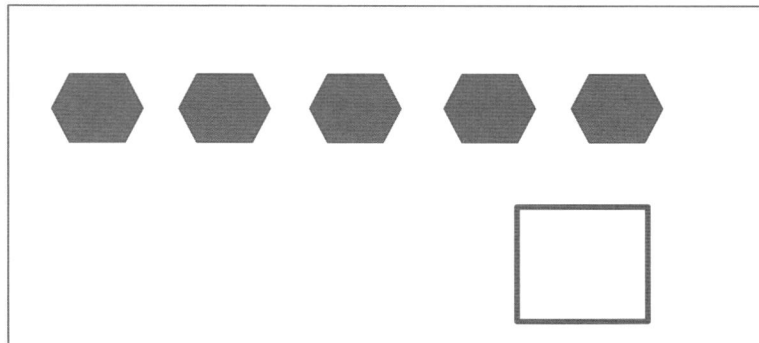

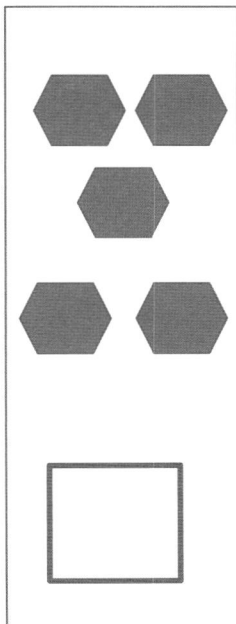

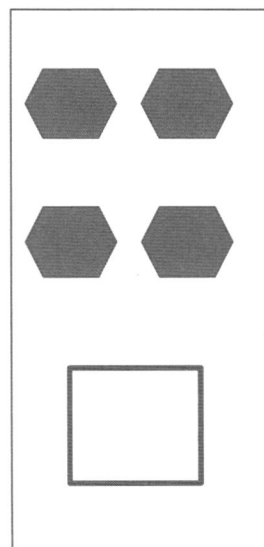

EUREKA MATH™   **Lesson 22:**   Arrange and strategize to count 8 beans in circular (around a cup) and scattered configurations. Write numeral 8. Find a path through the scatter set and compare paths with a partner.

Date: 6/6/14

Name _____   Date _____

Color 5 ladybugs. Color the remaining ladybugs a different color.
Count all the ladybugs. Write how many in the box.

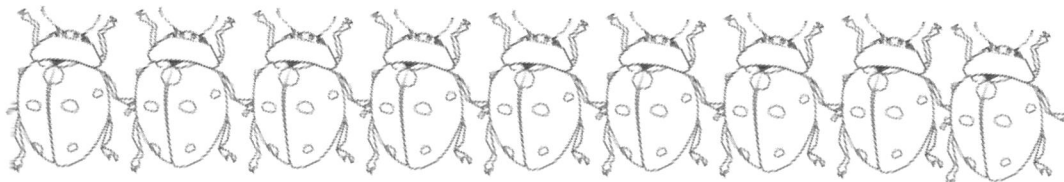

Color 5 diamonds. Color the remaining diamonds a different color.
Count all the diamonds. Write how many in the box.

Draw 4 more circles.
Count all the circles. Write how many in the box.

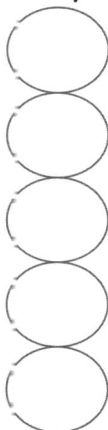

Make 9 dots. Circle a group of 5 dots.

EUREKA MATH

Lesson 23:   Organize and count 9 varied geometric objects in linear and array
(3 threes) configurations. Place objects on 5-group mat.
Match with numeral 9.
Date: 6/6/14

100

Color 3 ladybugs. Count all the ladybugs. Write how many in the box.

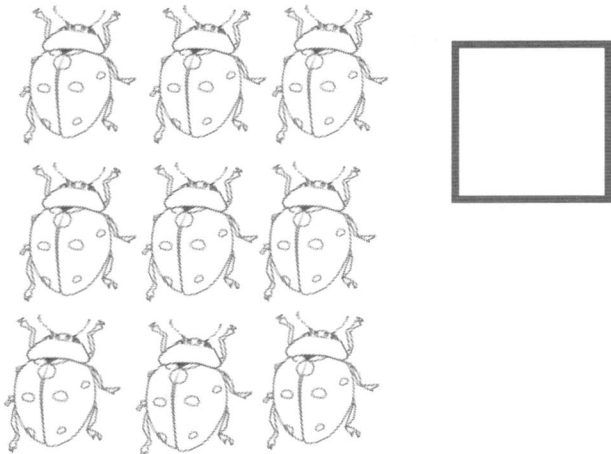

Color 3 rectangles. Count all the rectangles. Write how many in the box.

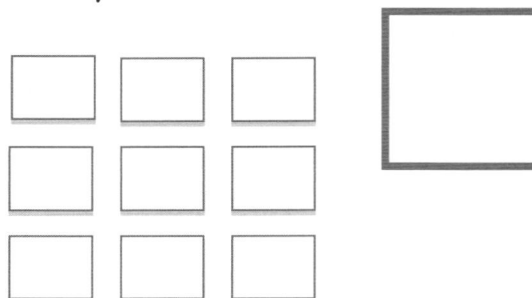

Draw 2 circles to finish the last row to make 9. Color to show the rows. Write how many in the box.

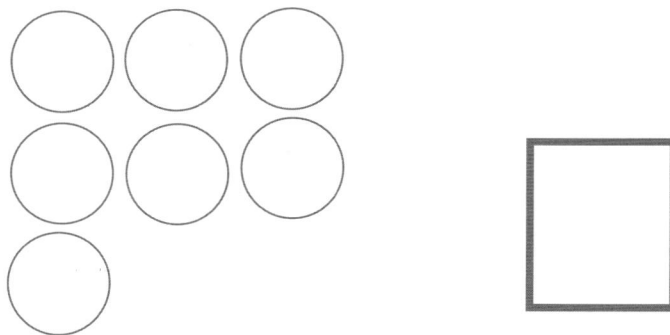

EUREKA MATH

Lesson 23: Organize and count 9 varied geometric objects in linear and array (3 threes) configurations. Place objects on 5-group mat. Match with numeral 9.

Date: 6/6/14

101

Name _____   Date _____

Color 5. Count how many shapes in all. Write the number in the box.

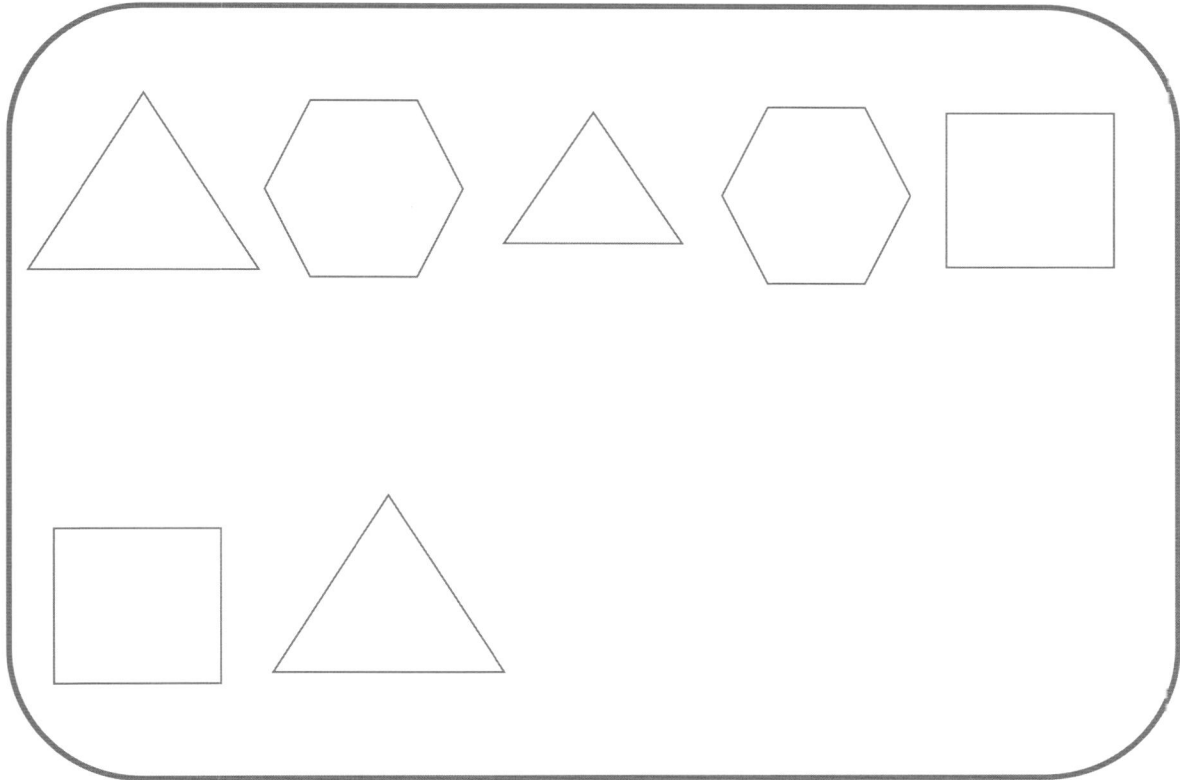

EUREKA
MATH™

Lesson 23:   Organize and count 9 varied geometric objects in linear and array
              (3 threes) configurations. Place objects on 5-group mat.
              Match with numeral 9.
Date: 6/6/14

102

Count how many dots. Write the number in the box.

Lesson 23:   Organize and count 9 varied geometric objects in linear and array
               (3 threes) configurations. Place objects on 5-group mat.
               Match with numeral 9.
Date: 6/6/14

103

Name _____ Date _____

Color 9 shapes.

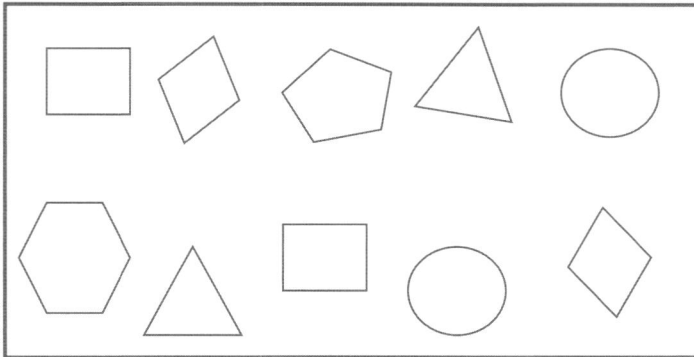

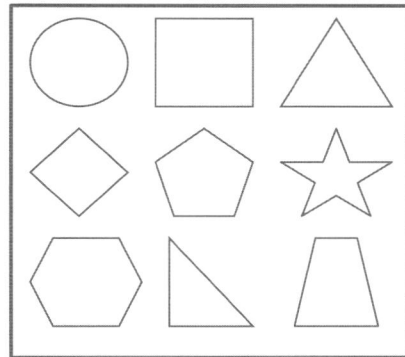

Color 9 shapes.

Draw 9 shapes.

Draw 9 shapes a different way.

EUREKA MATH™

Lesson 23:   Organize and count 9 varied geometric objects in linear and array (3 threes) configurations. Place objects on 5-group mat. Match with numeral 9.

Date: 6/6/14

104

Name _____   Date _____

Put this page into your personal white boards. Practice. When you are ready, use your pencil to write the numbers on the paper.

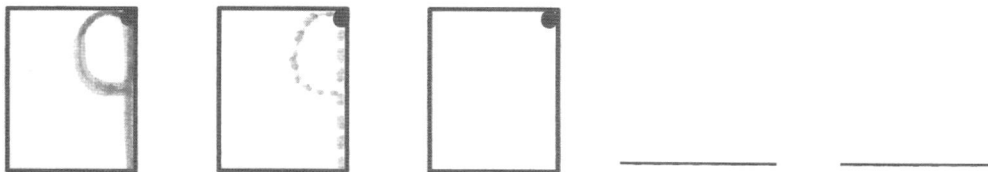

Color 9 happy faces.

Circle a different group of 9 happy faces.

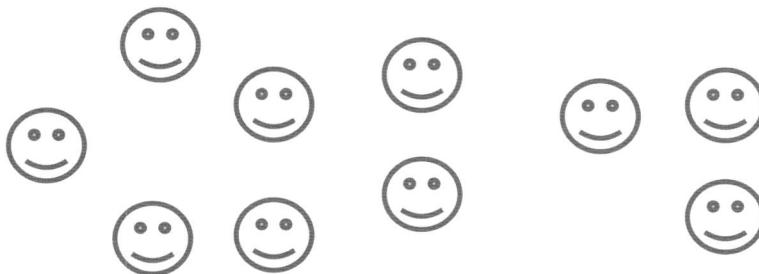

EUREKA MATH | **Lesson 24:** Strategize to count 9 objects in circular (around a paper plate) and scattered configurations printed on paper. Write numeral 9. Represent a path through the scatter count with pencil. Number each object.

Date: 6/6/14

105

Name _____ Date _____

Draw lines to connect the circles starting at 1.

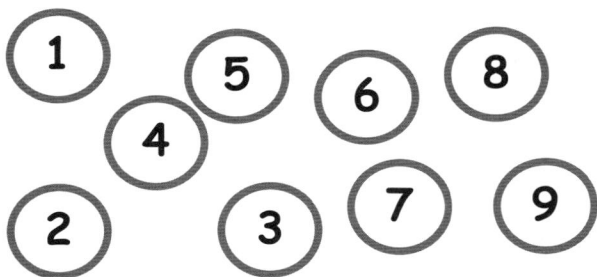

1 5 8
   4 6
2   3 7 9

Number the dots 1–9 in a different way. Connect the circles with lines.

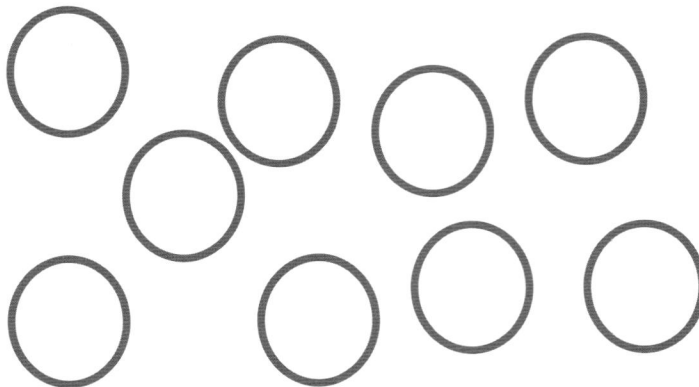

With your pencil, number the objects from 1 to 9 to show how you count the stars and objects. Write the total number of objects in the boxes.

EUREKA MATH

Lesson 24: Strategize to count 9 objects in circular (around a paper plate) and scattered configurations printed on paper. Write numeral 9. Represent a path through the scatter count with pencil. Number each object.

Date: 6/6/14

106

© 2014 Common Core, Inc. All rights reserved. commoncore.org

Count the dots.
Write the number.

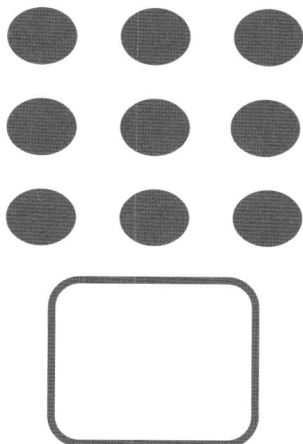

Count the dots. Write the number.
Circle a group of 5.

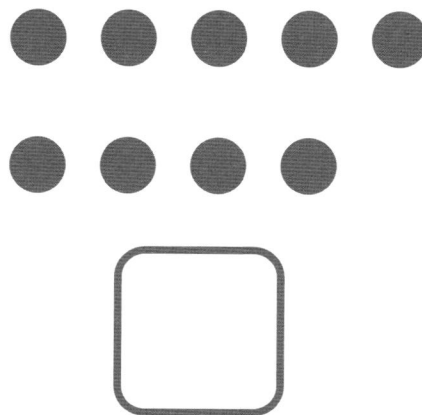

Draw more dots to make 9 in a circle.
Number the dots from 1 to 9.

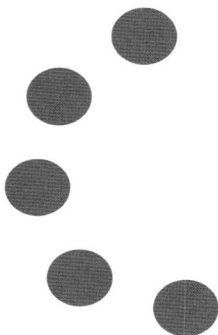

Count the dots. Circle 9 of them.
Within your 9, circle a group of 5.

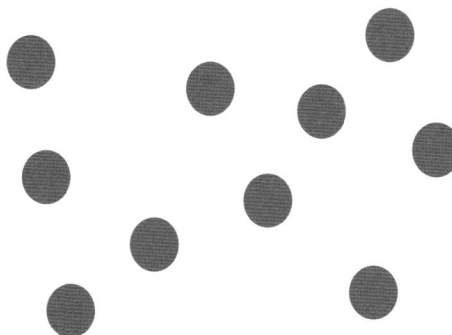

EUREKA MATH

Lesson 24: Strategize to count 9 objects in circular (around a paper plate) and scattered configurations printed on paper. Write numeral 9. Represent a path through the scatter count with pencil. Number each object.

Date: 6/6/14

107

Name _____   Date _____

Color 9 shapes.

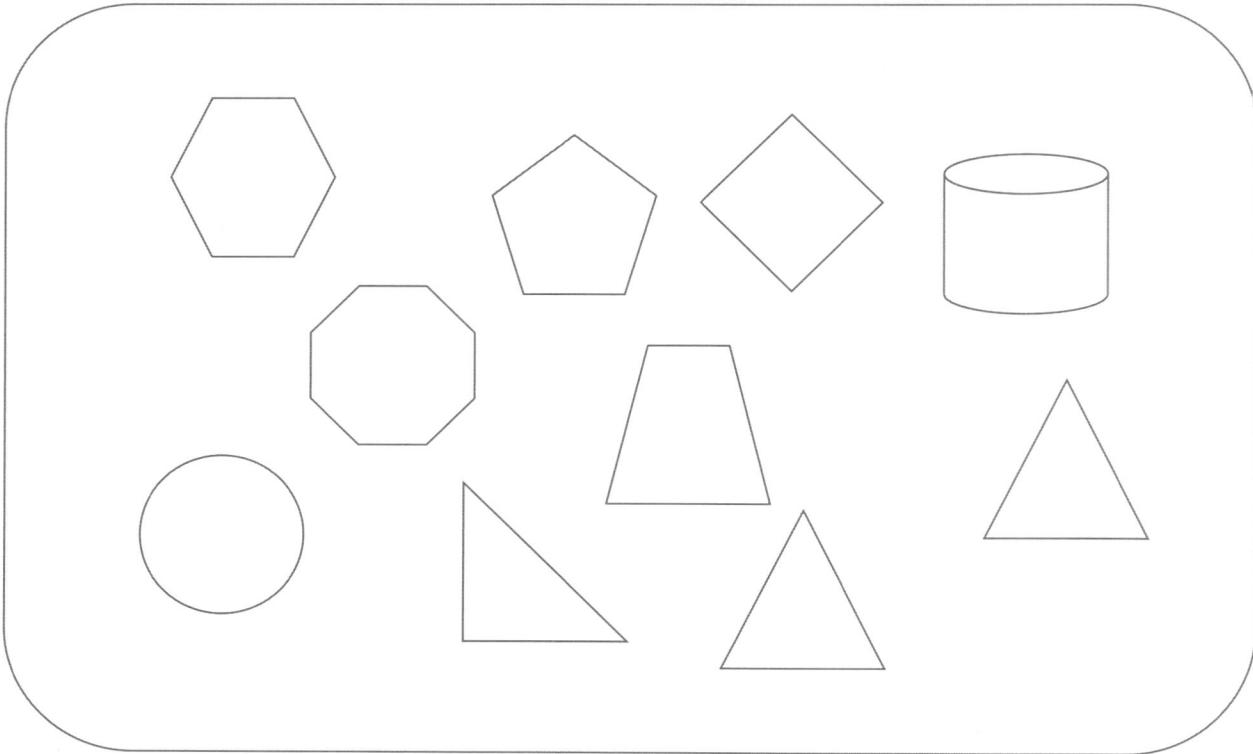

Draw 9 beans on the plate.

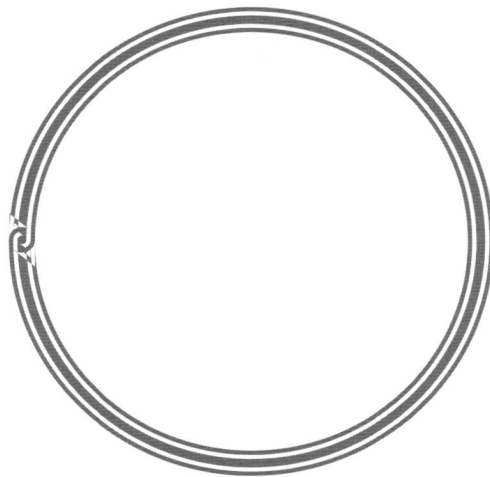

EUREKA MATH

Lesson 24:

Strategize to count 9 objects in circular (around a paper plate) and scattered configurations printed on paper. Write numeral 9. Represent a path through the scatter count with pencil. Number each object.

Date: 6/6/14

108

© 2014 Common Core, Inc. All rights reserved. commoncore.org

Name _____   Date _____

Number the circles from 1 to 9.   Color 9 circles.

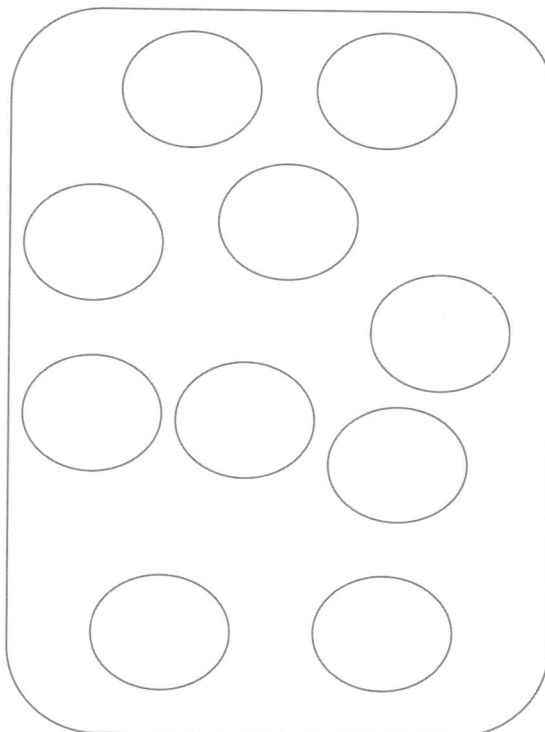

Draw 9 beads on the bracelet.   Count. Write the number in the box.

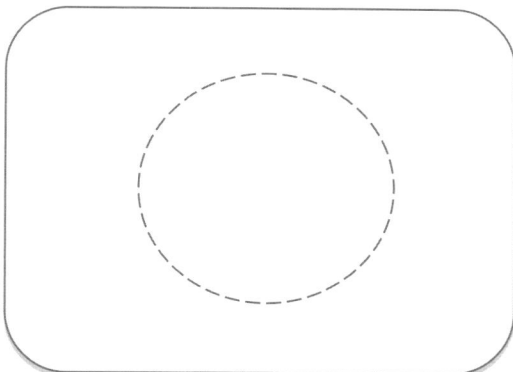

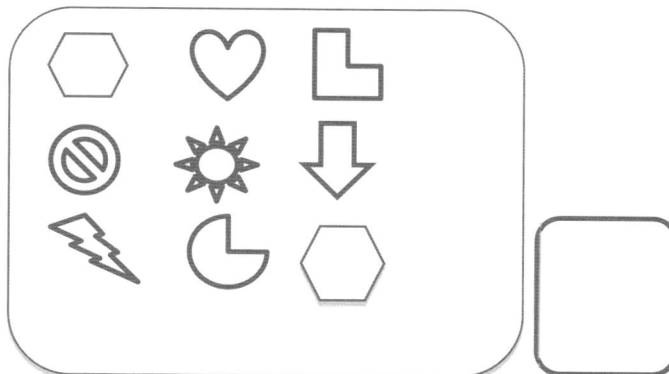

EUREKA MATH   **Lesson 24:**   Strategize to count 9 objects in circular (around a paper plate) and
scattered configurations printed on paper. Write numeral 9.
Represent a path through the scatter count with pencil. Number each
object.   109
Date: 6/6/14
© 2014 Common Core, Inc. All rights reserved. commoncore.org

blank 5-group

EUREKA
MATH™

Lesson 25: Count 10 objects in linear and array configurations (2 fives). Match
with numeral 10. Place on the 5-group mat. Dialogue about
9 and 10. Write numeral 10.

Date: 6/6/14

110

Name _____   Date _____

Color 5 ladybugs in a row. Color the remaining ladybugs a different color. Count all the ladybugs. Write how many in the box.

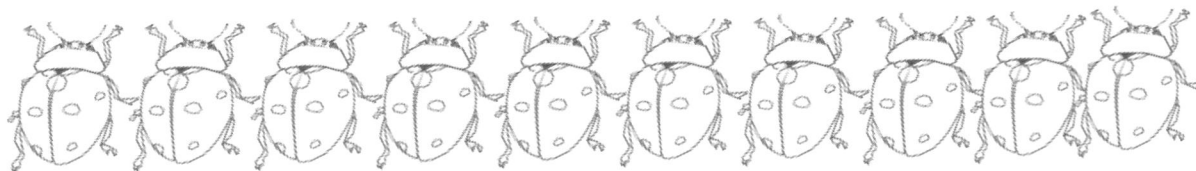

Color 5 diamonds in a row. Color the remaining diamonds a different color. Count all the diamonds. Write how many in the box.

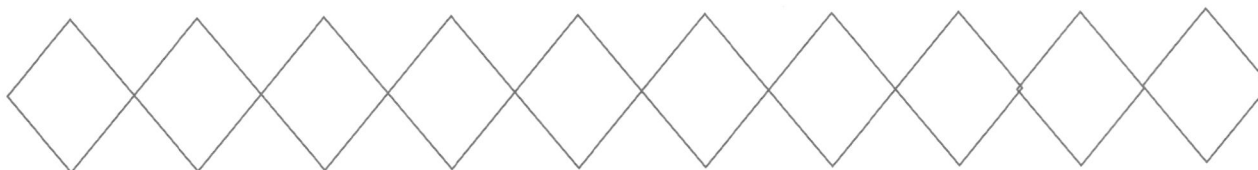

Color 5 circles. Then, draw 5 circles to the right. Count all the circles. Write how many in the box.

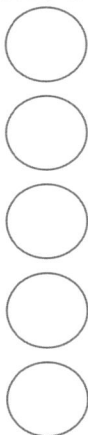

Color 5 circles. Then, draw 5 circles below. Count all the circles. Write how many in the box.

EUREKA MATH

Lesson 25:   Count 10 objects in linear and array configurations (2 fives). Match with numeral 10. Place on the 5-group mat. Dialogue about 9 and 10. Write numeral 10.

Date: 6/6/14

111

Color 5 ladybugs. Count all the ladybugs. Write how many in the box.

Color 5 squares. Count all the squares. Write how many in the box.

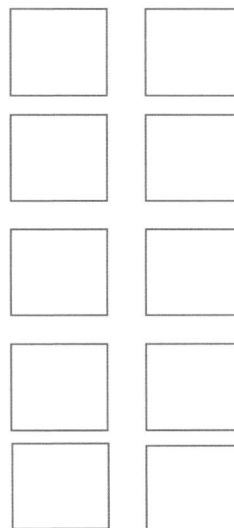

Color 5 circles. Draw 4 circles to finish the row. Color the bottom 5 a different color. Write how many circles in all in the box.

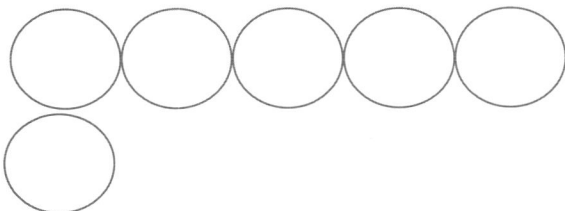

EUREKA MATH™

Lesson 25:   Count 10 objects in linear and array configurations (2 fives). Match with numeral 10. Place on the 5-group mat. Dialogue about 9 and 10. Write numeral 10.
Date: 6/6/14

112

Name _____   Date _____

Draw 5 more circles. How many are there now? Write how many in the box.

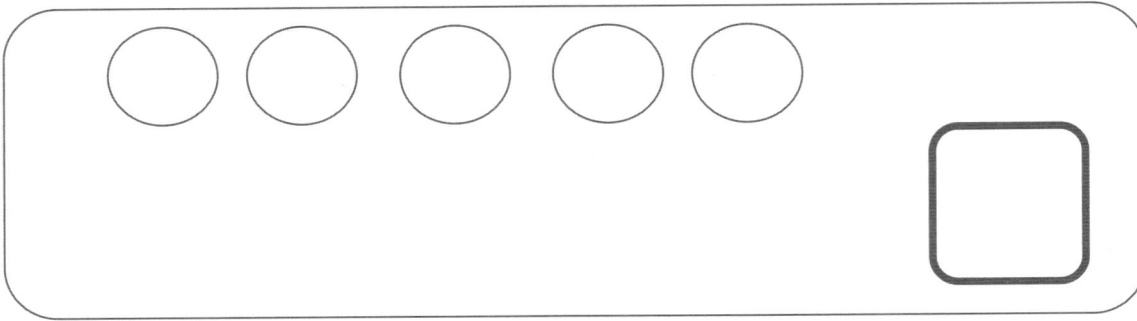

Color 5 blocks blue. Color 5 blocks green. Write how many in the box.

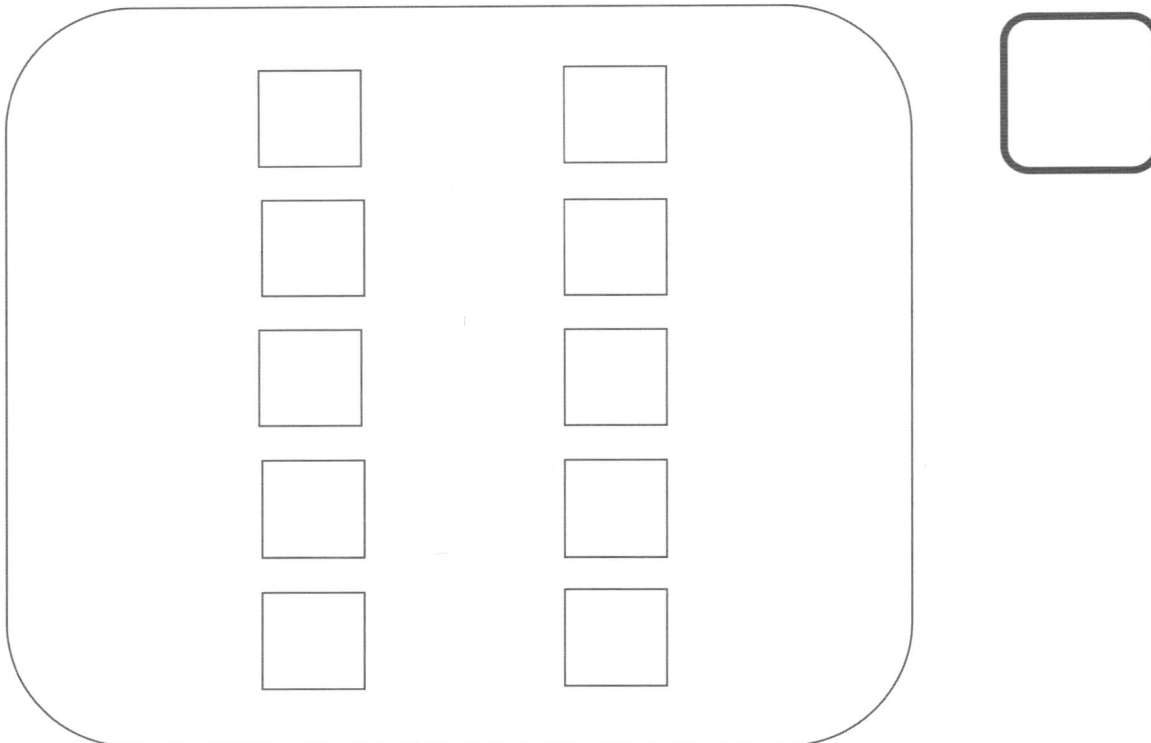

EUREKA MATH

Lesson 25:   Count 10 objects in linear and array configurations (2 fives). Match with numeral 10. Place on the 5-group mat. Dialogue about 9 and 10. Write numeral 10.

Date: 6/6/14

113

Name _____   Date _____

Color 9 squares. Color 1 square a different color.

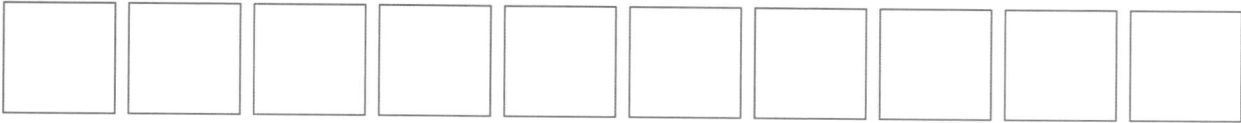

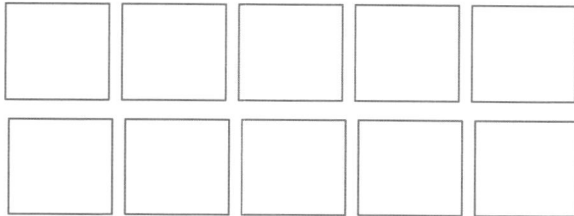

Color 9 squares. Color 1 square a different color.

Color 5 squares. Color 5 squares a different color.

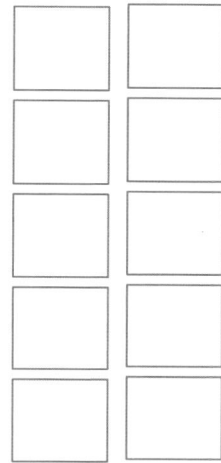

Draw 10 circles in a line. Color 5 circles red. Color 5 circles blue.

Draw 5 circles under the row of circles. Color 5 circles red. Color 5 circles blue.

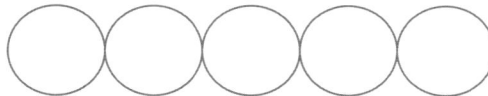

EUREKA MATH™

Lesson 25:   Count 10 objects in linear and array configurations (2 fives). Match with numeral 10. Place on the 5-group mat. Dialogue about 9 and 10. Write numeral 10.

Date: 6/6/14

Name _____   Date _____

Put this page into your personal white boards. Practice. When you are ready, write your numbers in pencil on the paper.

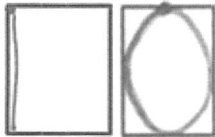      _____

      _____

      _____

Name _____   Date _____

Draw 10 circles in a row. Color the first 5 yellow, the second 5 blue. Write
how many circles in the box.

Draw 5 circles in the gray part. Draw 5 circles in the white part.
Write how many circles in the box.

| Draw two towers of 5 next to each other. Color 1 tower red and the other tower orange. Circle groups of two. | Draw a row of 5 cubes. Draw another row of 5. Count. Write how many cubes. |

Draw a picture of your bracelet on the back of the paper.

EUREKA
MATH

Lesson 26:   Count 10 objects in linear and array configurations (2 fives). Match
              with numeral 10. Place on the 5-group mat. Dialogue about
              9 and 10. Write numeral 10.
Date: 6/6/14

116

© 2014 Common Core, Inc. All rights reserved. commoncore.org

Name _____   Date _____

Color 5 blocks red and 5 blocks green. How many blocks? Write how many in the box.

Color 5 blocks brown and 5 blocks yellow. How many blocks? Write how many in the box.

EUREKA MATH

Lesson 26: Count 10 objects in linear and array configurations (2 fives). Match with numeral 10. Place on the 5-group mat. Dialogue about 9 and 10. Write numeral 10.

Date: 6/6/14

Name _____   Date _____

Draw 5 triangles in a row. Draw another 5 triangles in a row under them.

How many triangles did you draw?

Write the number in the box.

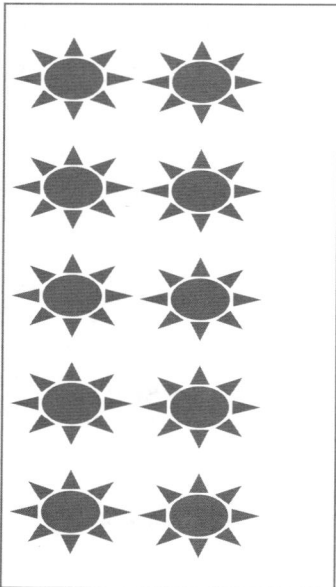

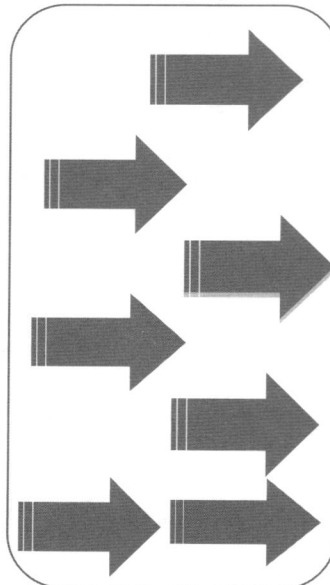

Write how many

in the box.

Write how many

in the box.

EUREKA MATH™

Lesson 26:   Count 10 objects in linear and array configurations (2 fives). Match with numeral 10. Place on the 5-group mat. Dialogue about 9 and 10. Write numeral 10.

Date: 6/6/14

118

Name _____   Date _____

Count the shapes, and write how many. Color the shape you counted first.

Draw 10 things. Color 5 of them. Color 5 things a different color.

Draw 10 circles. Color 5 circles. Color 5 circles a different color.

Show how many apples by drawing a path between them as you count.

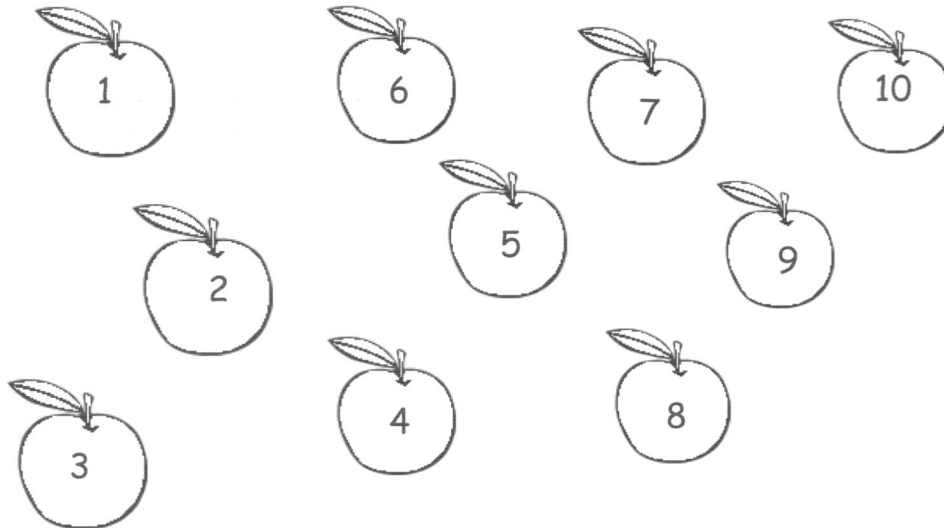

1   6   7   10

2   5   9

3   4   8

Write the numbers 1 to 10 in the apples a different way. When you count, draw a path connecting the apples.

Name _____   Date _____

Draw 10 beads on the bracelet.

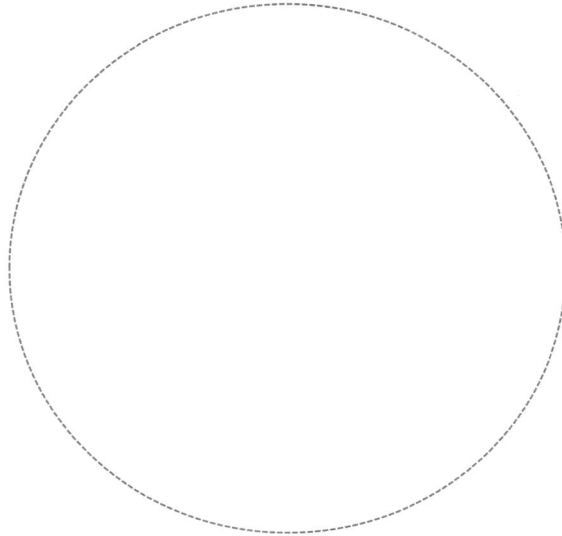

Count and write the numbers 1 to 10 in the ⬠ . Write how many in the box.

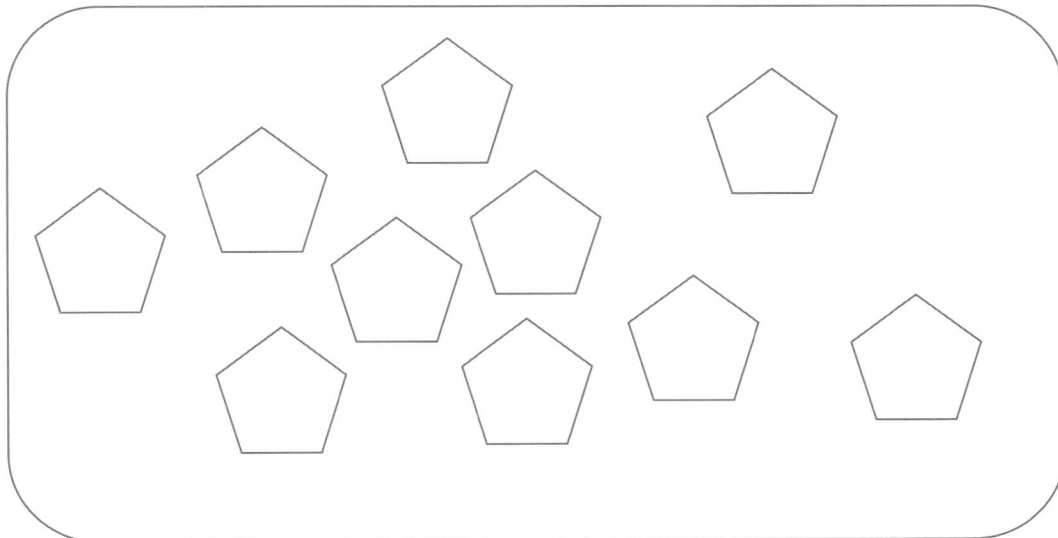

Name _____   Date _____

Draw enough  to make 10.

Draw enough to make 10.

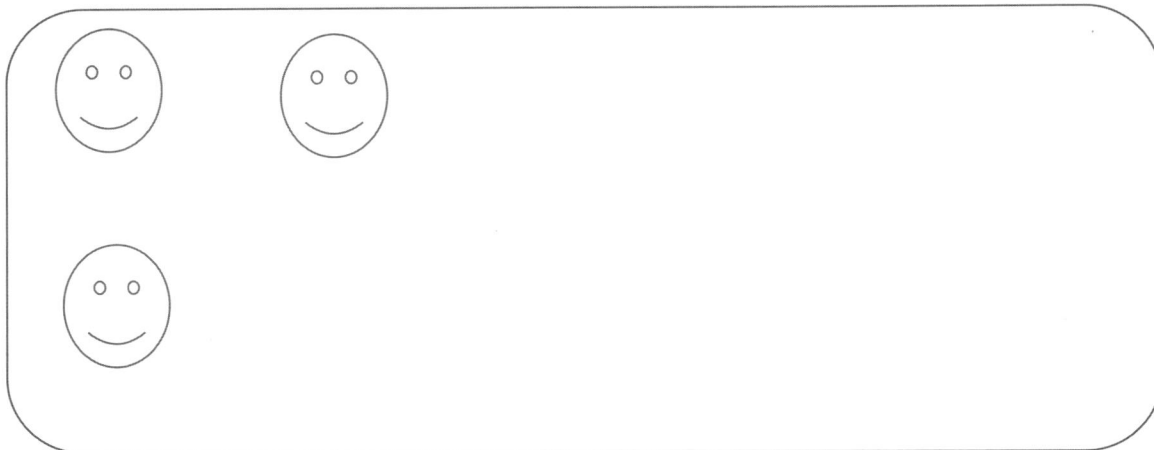

EUREKA
MATH™

Lesson 27:   Count 10 objects and move between all configurations.
Date: 6/6/14

122

Name _____   Date _____

Listen to my stories. Color the pictures to show what is happening. Write how many in the box.

Bobby picked 4 red flowers. Then, he picked 2 purple flowers. How many flowers did Bobby pick?

Janet went to the donut store. She bought 6 chocolate donuts and 3 strawberry donuts. How many donuts did she buy?

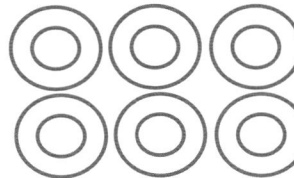

Some children were sitting in a circle. 4 of them were wearing green shirts. The rest were wearing yellow shirts. How many children were in the circle?

Jerry spilled his bag of marbles. Circle the group of grey marbles. Circle the group of black marbles. How many marbles were spilled?

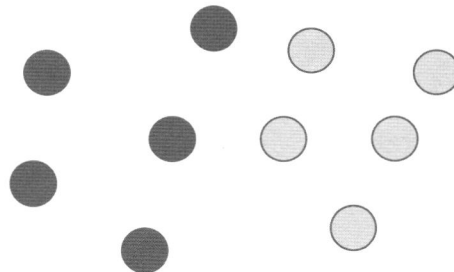

Make up a story about the bears. Color the bears to match the story.
Tell your story to a friend.

Make up a new story. Draw a picture to go with your story. Tell your
story to a friend.

Name _____   Date _____

How many ⬠ ?   Write how many in the box. ☐

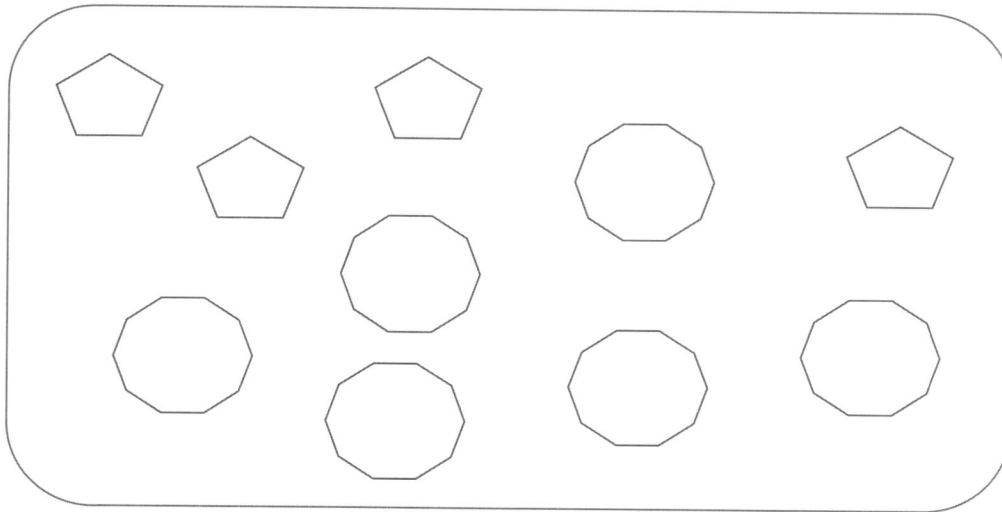

Draw 6 circles. Draw 4 triangles.   ☐

How many shapes did you draw? Write how many in the box.

Name _____   Date _____

Make up a story about 10 things in your house. Draw a picture to go with your story. Be ready to share your story at school tomorrow.

Make your own 5-group cards! Cut the cards out on the dotted lines. On one side, write the numbers from 1 to 10. On the other side, show the 5-group dot picture that goes with the number.

EUREKA
MATH

Lesson 29:   Order and match numeral and dot cards from 1 to 10. State 1 more
              than a given number.
Date: 6/6/14

131

© 2014 Common Core, Inc. All rights reserved. commoncore.org

piggy bank mat

EUREKA
MATH™

Lesson 29: Order and match numeral and dot cards from 1 to 10. State 1 more
than a given number.
Date: 6/6/14

132

Name _____   Date _____

Count and color the white squares red. Count all the cubes in each step.
Write the missing numbers below each step.

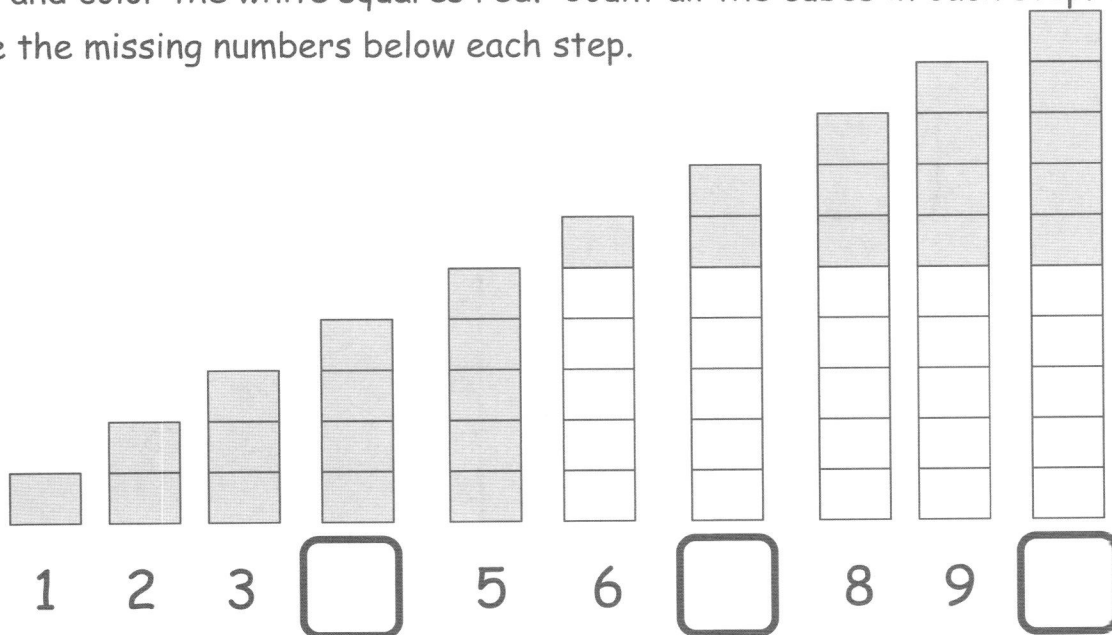

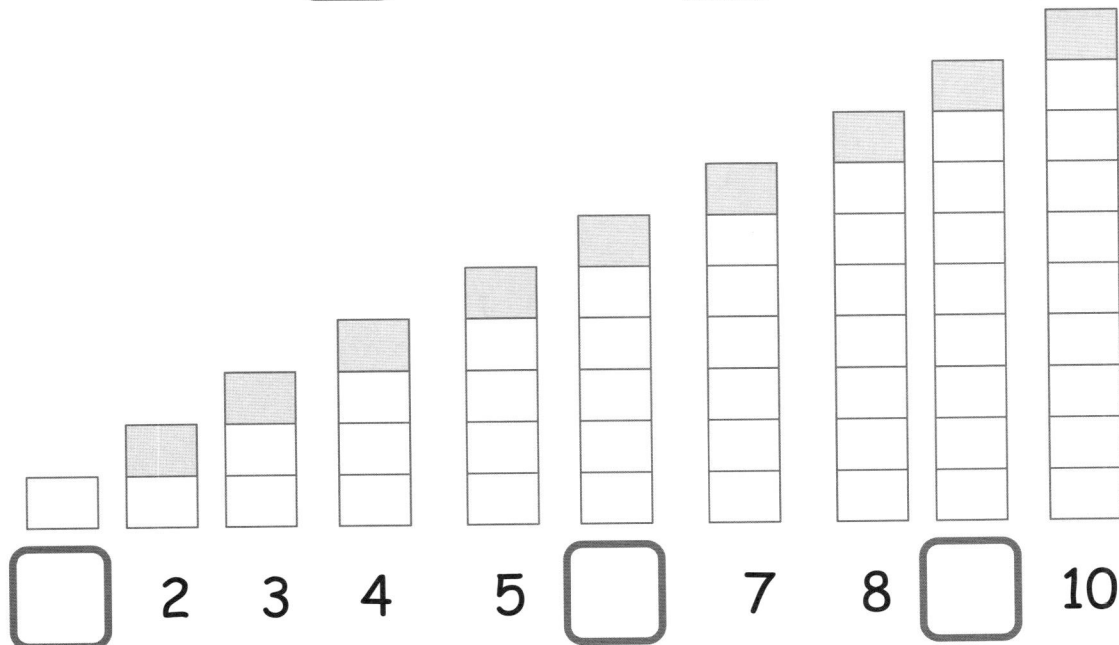

1   2   3   ☐   5   6   ☐   8   9   ☐

☐   2   3   4   5   ☐   7   8   ☐   10

Name _____ Date _____

Draw a stair that shows 1 more, and write the new number in the box.

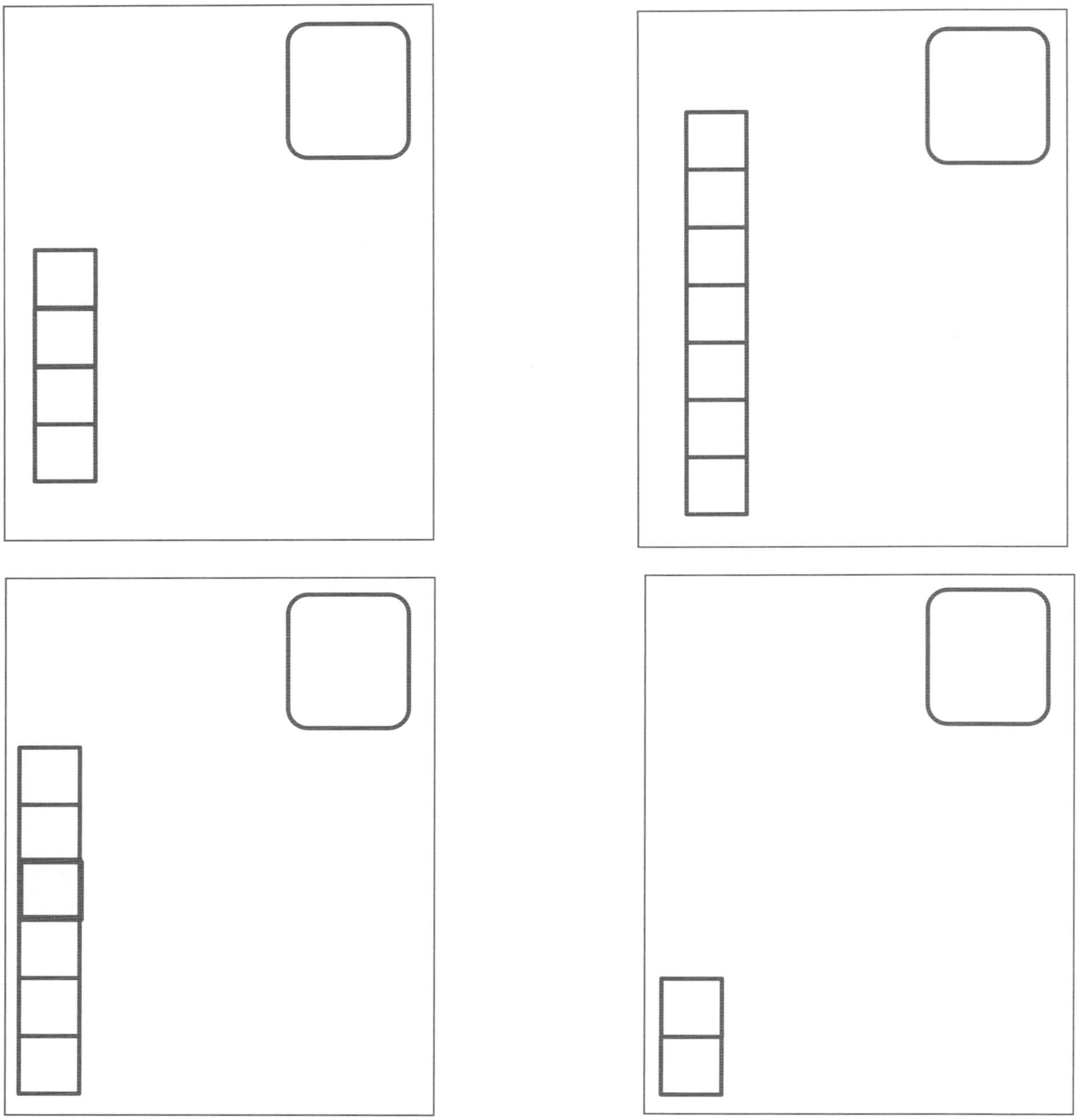

EUREKA MATH

Lesson 30: Make math stairs from 1 to 10 in cooperative groups.
Date: 6/6/14

134

Name _____ Date _____

Draw the missing stairs. Write the numbers below each step.

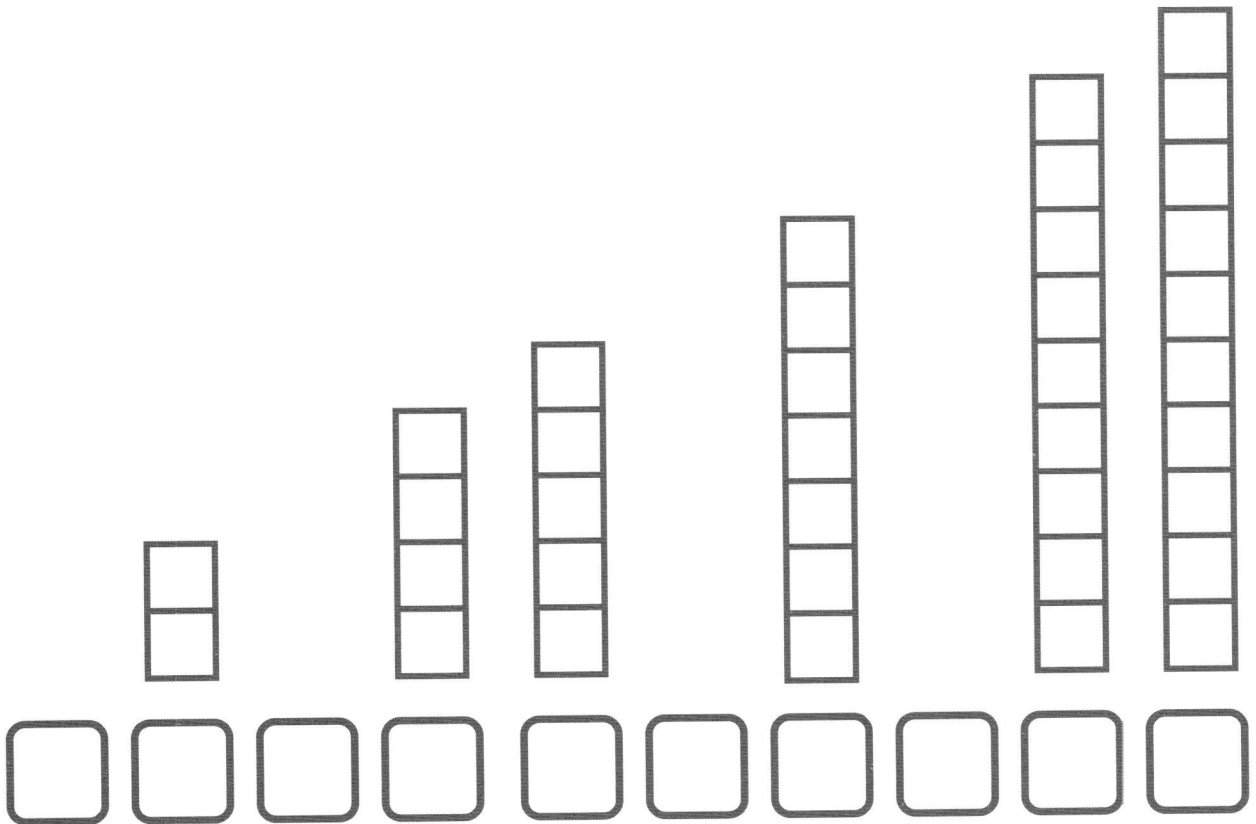

Ask someone to help you write about what you think baby bear will do now that you have helped him to get home. Use the back of this paper.

EUREKA MATH

Lesson 30: Make math stairs from 1 to 10 in cooperative groups.
Date: 6/6/14

135

Draw 1 more cube on each stair so the cubes match the number. Say as you draw, "1. One more is two. 2. One more is three."

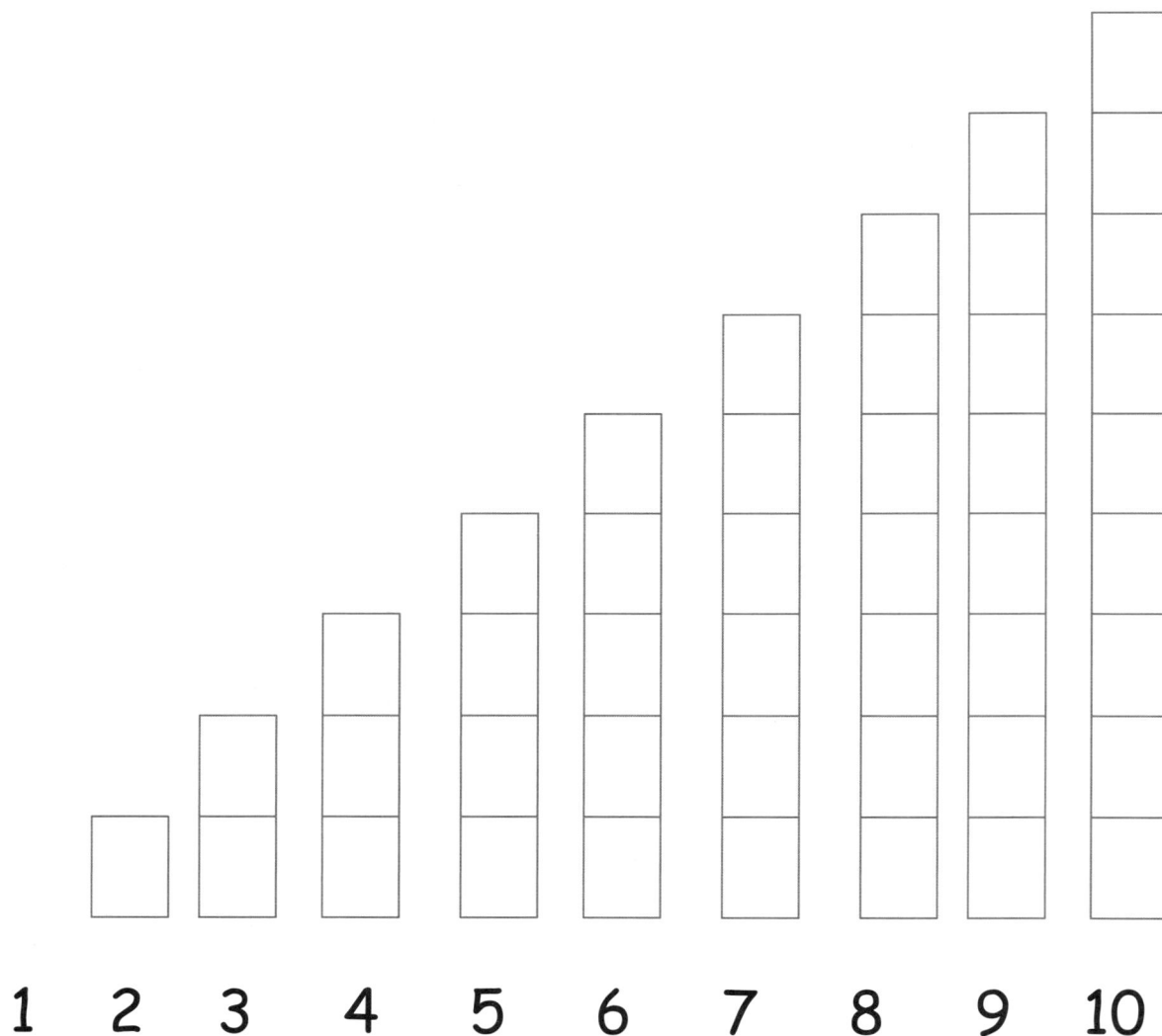

1   2   3   4   5   6   7   8   9   10

bears template

EUREKA MATH

Lesson 30:   Make math stairs from 1 to 10 in cooperative groups.
Date: 6/6/14

137

Name _____   Date _____

Color the empty circles orange and count. Count the gray circles, and write
how many gray circles in the box.

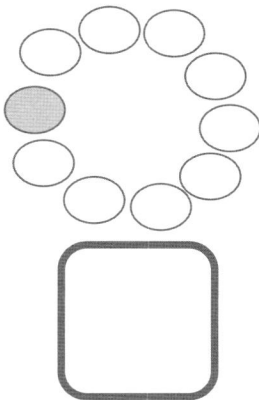

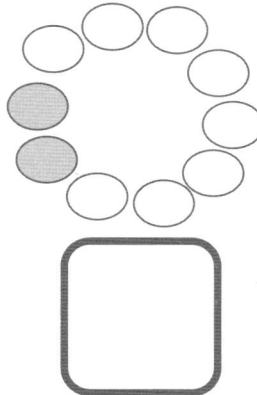

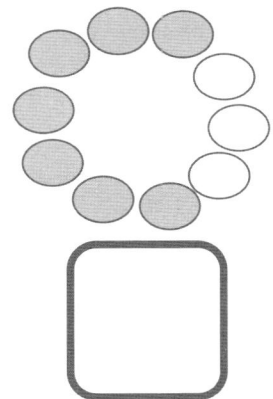

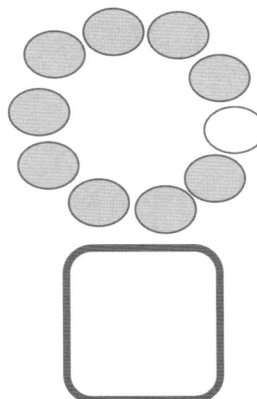

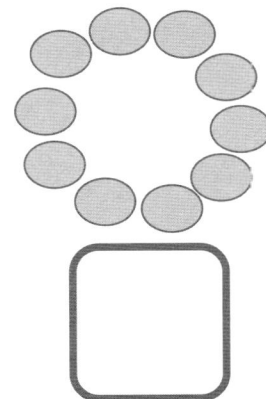

EUREKA
MATH

Lesson 31:   Arrange, analyze, and draw 1 more up to 10 in configurations other
than towers.
Date: 6/6/14

138

Count the white circles and color them blue. Draw 1 more and count all the circles. Write how many.

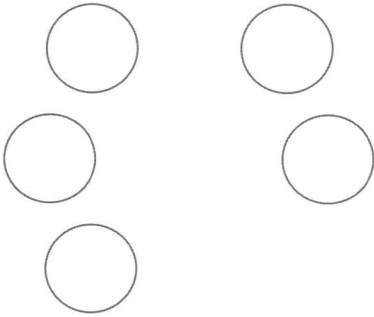

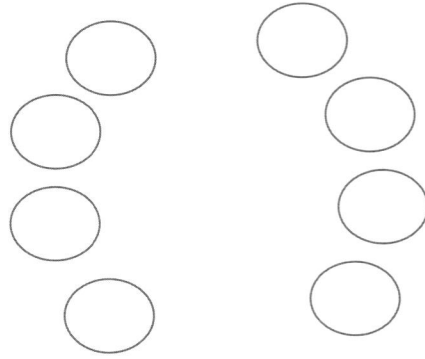

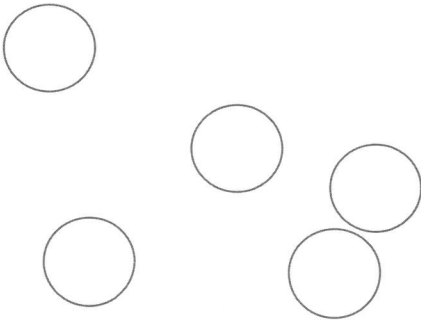

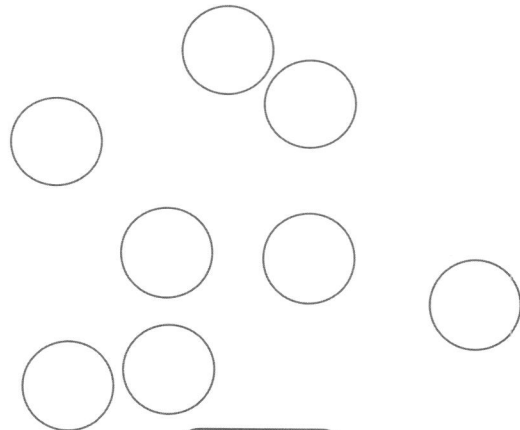

Name _____   Date _____

| Color the stars blue. Draw 1 more star. Color it blue, and write how many. | Color the happy faces red. Draw 1 more happy face. Color it red, and write how many. |

EUREKA MATH™   **Lesson 31:**   Arrange, analyze, and draw 1 more up to 10 in configurations other than towers.
Date: 6/6/14

Name _____　Date _____

Draw one more square. Color all the squares and write how many.

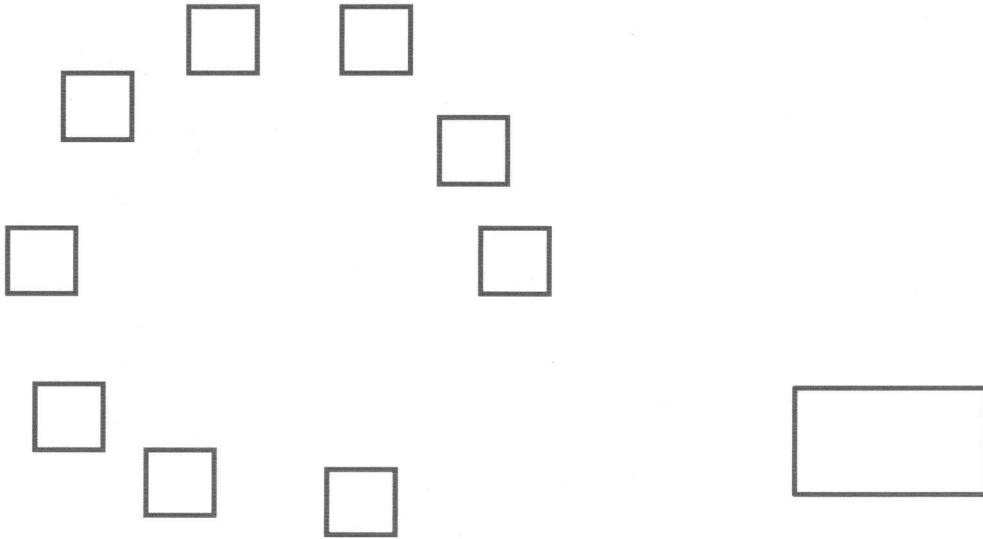

Draw one more cloud. Color all the clouds and write how many.

Name _____   Date _____

Draw 1 more, and write how many in the box.

	How many?		How many?
△		▪▪ ▪ ▪ ▪ ▪▪	
☐ ☐ ☐		OOOO OOO	
⬭ ⬭ ⬭ ⬭		△△△ △△△	
△ △ △ △		☐ ☐ ☐ ☐ ☐ ☐	
▪ ▪ ▪ ▪		⬭⬭⬭⬭⬭ ⬭⬭⬭	
⬭⬭⬭⬭⬭⬭		△△△△△ △△△	
△△△△△△		▪ ▪ ▪ ▪ / ▪ ▪ ▪ ▪	
● / ●●●●●		△ △ △ / △△ △△ △△	

_____

draw 1 more

EUREKA MATH

Lesson 32: Arrange, analyze, and draw sequences of quantities of 1 more, beginning with numbers other than 1.
Date: 6/6/14

142

Name _____   Date _____

Draw and write the number of the missing steps.

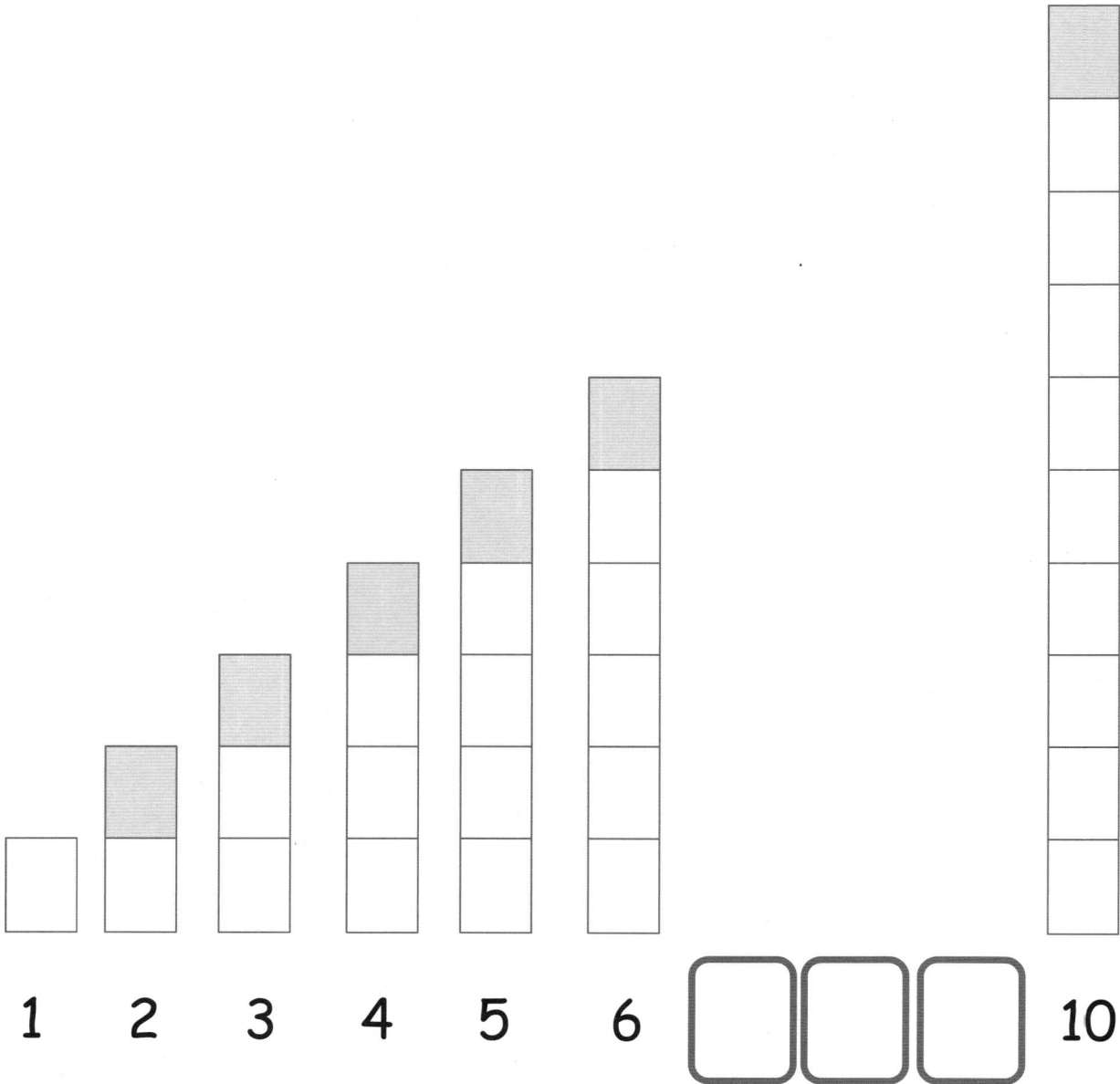

1   2   3   4   5   6   ☐   ☐   ☐   10

EUREKA MATH   **Lesson 32:**   Arrange, analyze, and draw sequences of quantities of 1 more, beginning with numbers other than 1.
Date: 6/6/14

143

Write the missing number. Draw objects to show the numbers.

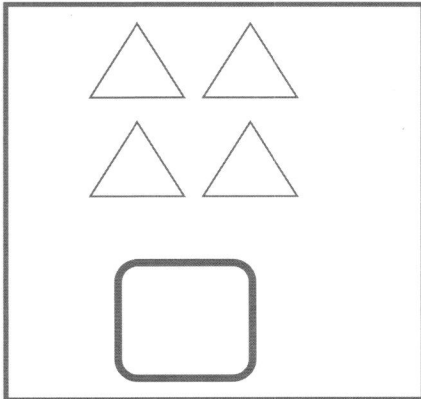

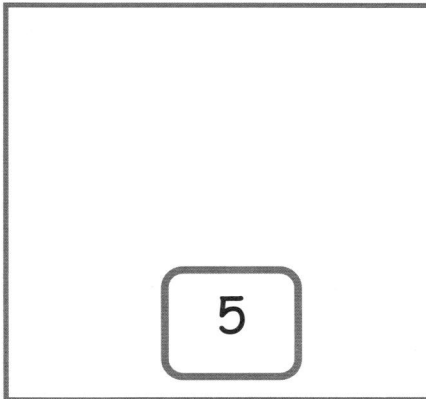

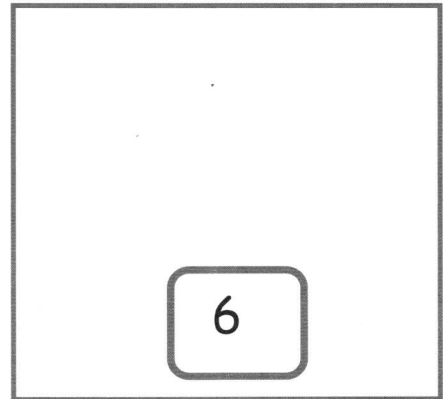

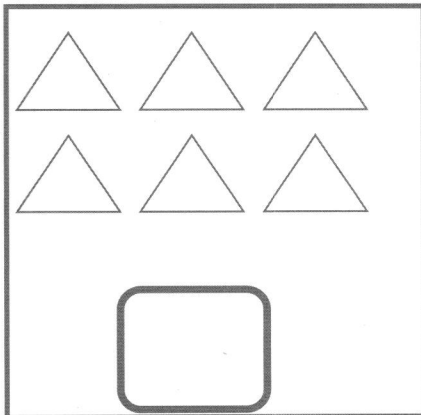

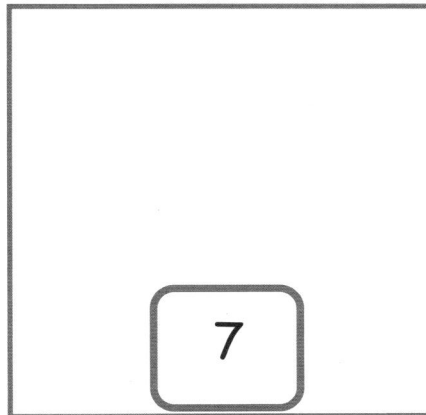

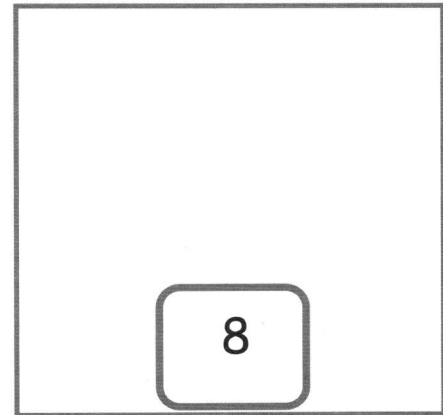

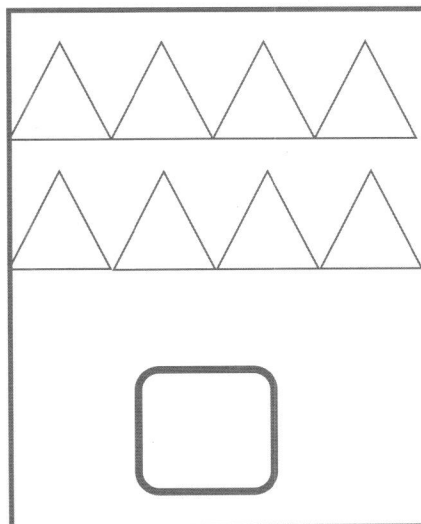

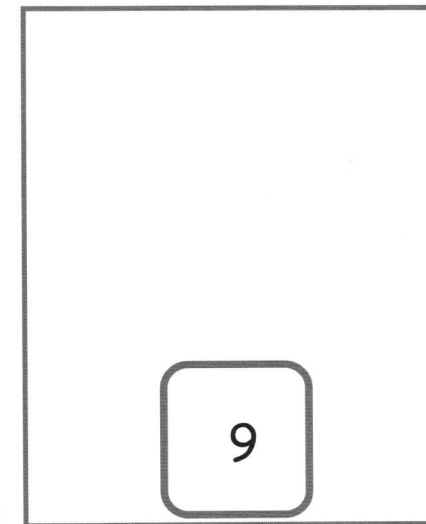

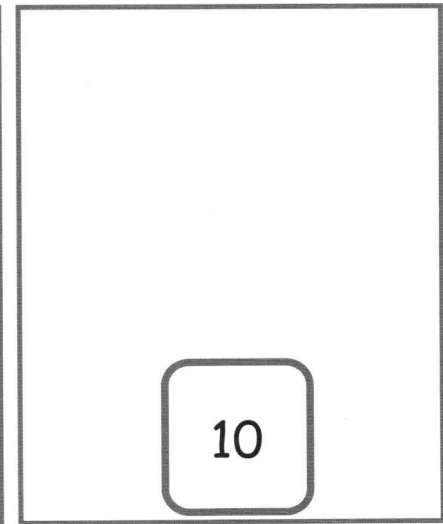

EUREKA
MATH

Lesson 32:   Arrange, analyze, and draw sequences of quantities of 1 more,
              beginning with numbers other than 1.
Date: 6/6/14

144

© 2014 Common Core, Inc. All rights reserved. commoncore.org

Name _____   Date _____

Write the missing numbers.

3, _____, _____, 6, 7, _____, _____, _____

Draw 1 more apple each time.

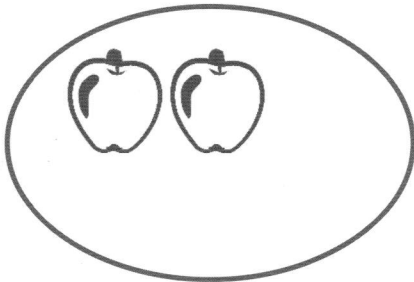

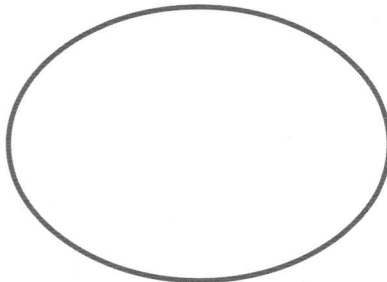

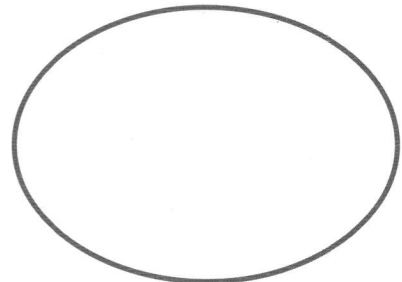

EUREKA MATH | **Lesson 32:** Arrange, analyze, and draw sequences of quantities of 1 more, beginning with numbers other than 1.
Date: 6/6/14

145

Name _____   Date _____

Write the missing numbers.

___, 2, ___, ___, ___, 6, 7, ___, ___, 10

Draw X's or O's to show 1 more.

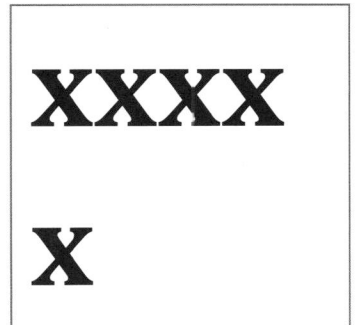

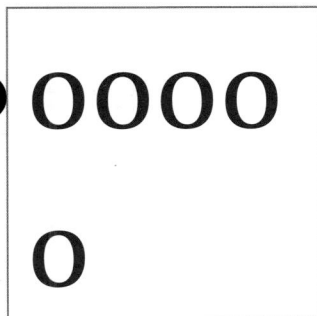

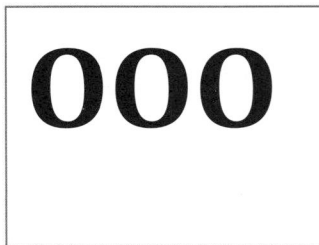

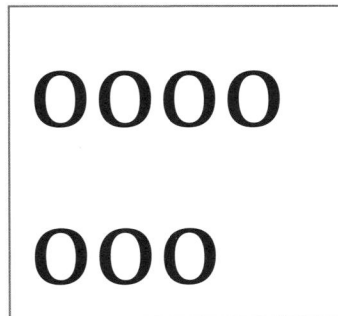

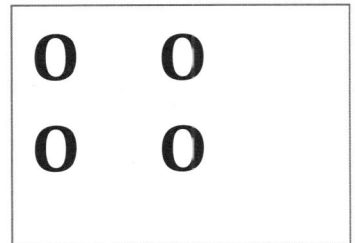

XX	X		XXXX X
OOOO O	OOO	OOOO OOO	O O O O

Tell someone a story about "1 more...and then 1 more." Draw a picture about your story.

EUREKA MATH   Lesson 32: Arrange, analyze, and draw sequences of quantities of 1 more, beginning with numbers other than 1.
Date: 6/6/14   146

© 2014 Common Core, Inc. All rights reserved. commoncore.org

Name _____   Date _____

Count the dots. Write how many in the circle. Draw the same number of dots below the circle, but going up and down instead of across. The number 6 is done for you.

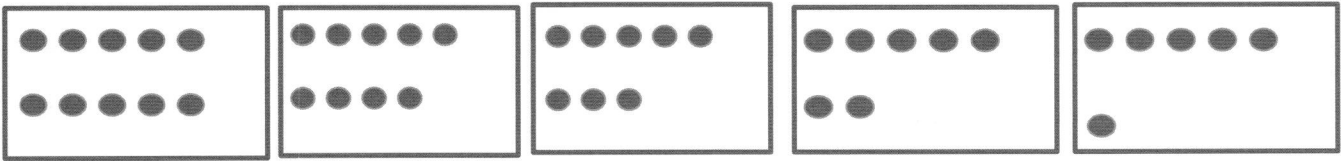

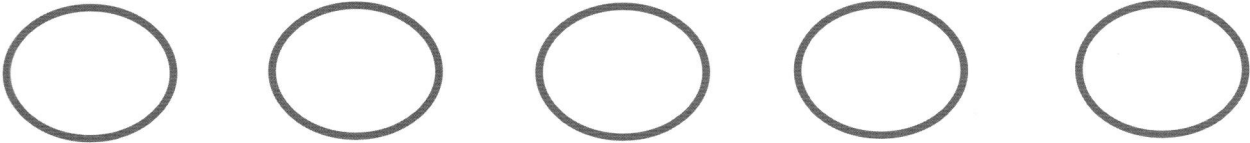

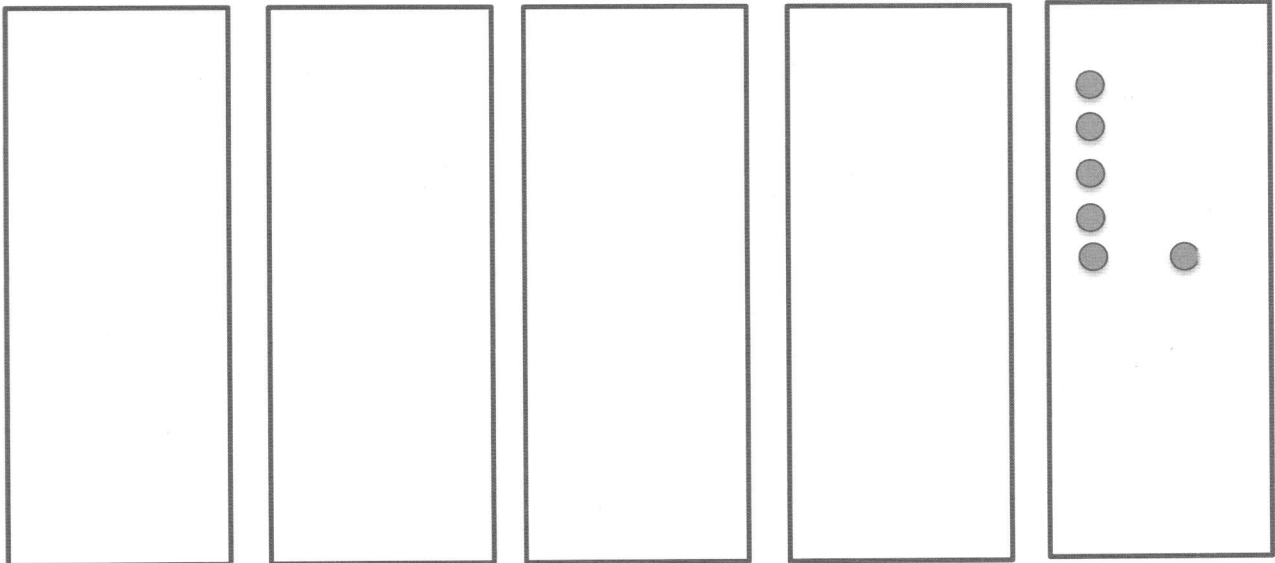

EUREKA MATH™   Lesson 33:   Order quantities from 10 to 1 and match numerals.
Date: 6/6/14

Count the dots. Write how many in the circle. Draw the same number of dots below the circle, but go up and down instead of across. The number 4 is done for you.

EUREKA
MATH

Lesson 33:   Order quantities from 10 to 1 and match numerals.
Date: 6/6/14

Count the balloons. Cross out 1 balloon. Count and write how many balloons are left in the box.

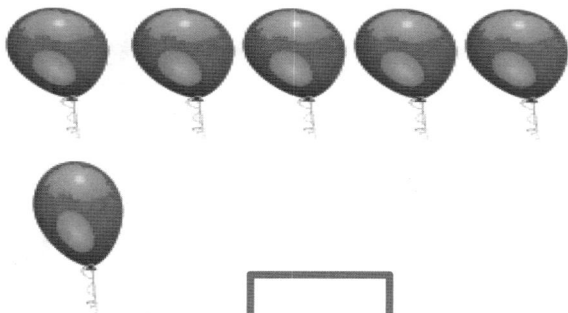

Count the basketballs. Cross out 1 basketball. Count and write how many basketballs are left in the box.

Count the balloons. Cross out 1 balloon. Count and write how many balloons are left in the box.

Count the basketballs. Cross out 1 basketball. Count and write how many basketballs are left in the box.

EUREKA MATH™

Lesson 33: Order quantities from 10 to 1 and match numerals.
Date: 6/6/14

149

Name _____   Date _____

Draw a line to match the picture to its number.

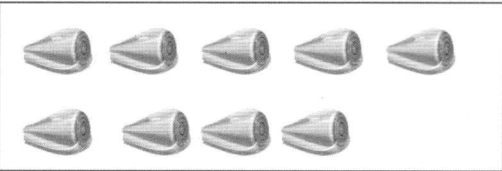

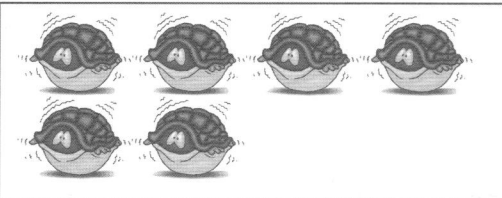

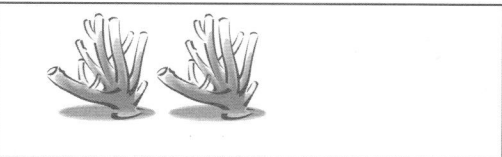

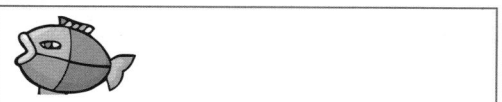

8

6

9

10

7

1

4

2

3

5

Lesson 33:   Order quantities from 10 to 1 and match numerals.
Date: 6/6/14

Make 5-group Cards.

Cut the cards out on the dotted lines. On one side, write the numbers from 1-10. On the other side, show the 5-group dot picture that goes with the number. Mix up your cards, and practice putting them in order in the "1 less" way.

EUREKA MATH

Lesson 33: Order quantities from 10 to 1 and match numerals.
Date: 6/6/14

151

Name _____   Date _____

Read the story to the students. Have students cut out the robots. Read the story again as the students glue the robots on the 5-group mat.

10 robots were playing in a circle. 1 robot's mom called, and he had to go home. 10. One less is nine. 9 robots were playing in a circle. 1 robot's mom called, and he had to go home. 9. One less is eight. 8 robots were playing in a circle. 1 robot's mom called, and he had to go home. 8. One less is seven. 7 robots were playing in a circle. 1 robot's mom called, and he had to go home. 7. One less is six. 6 robots were playing in a circle. 1 robot's mom called, and he had to go home. 6. One less is five. 5 robots were playing in a circle. 1 robot's mom called, and he had to go home. 5. One less is four. 4 robots were playing in a circle. 1 robot's mom called, and he had to go home. 4. One less is three. 3 robots were playing in a circle. 1 robot's mom called, and he had to go home. 3. One less is two. 2 robots were playing in a circle. 1 robot's mom called, and he had to go home. 2. One less is one. And, he played happily ever after!

1	2	3	4	5
6	7	8	9	10

EUREKA
MATH™

Lesson 34: Count down from 10 to 1, and state 1 less than a given number.
Date: 6/6/14

153

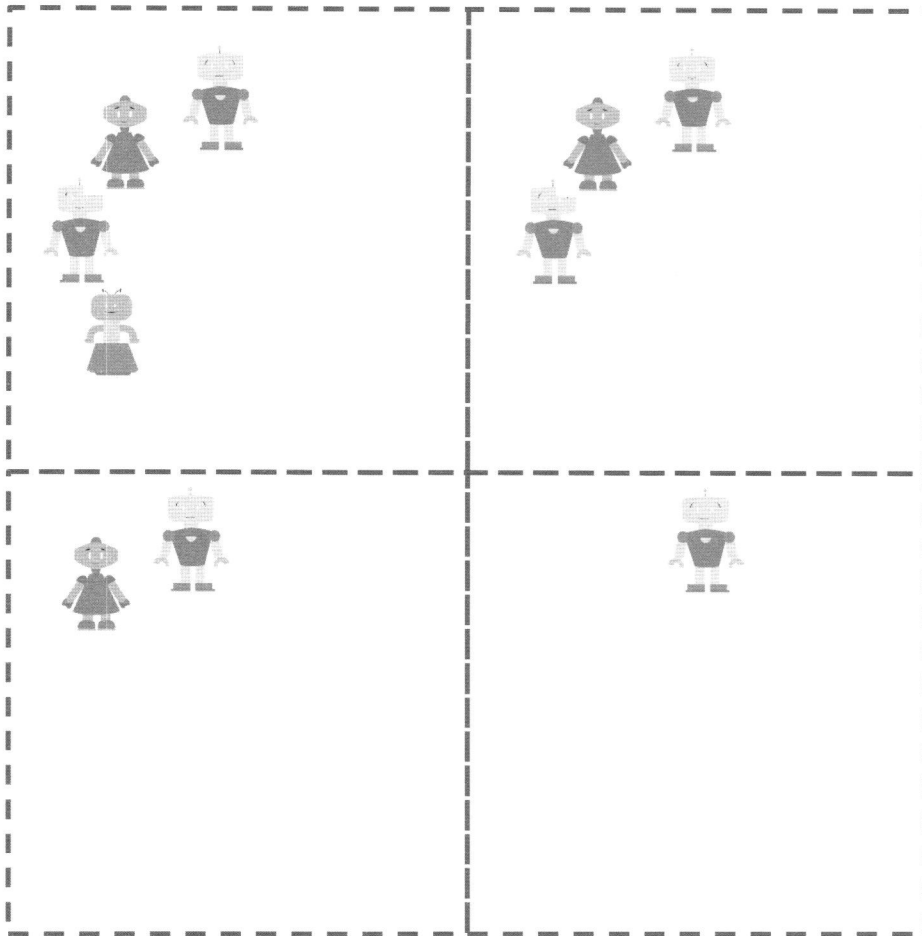

EUREKA
MATH

Lesson 34:   Count down from 10 to 1, and state 1 less than a given number.
Date: 6/6/14

154

Name _____   Date _____

Count and write the number of apples. Color only the group of apples that is 1 less.

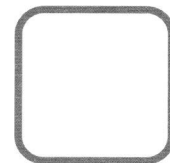

Count and write the number of hearts. Color only the group of hearts that is 1 less.

Name _____   Date _____

Count and color the triangles. Draw a group of triangles that is 1 less.
Write how many you drew.

Count and color the pears. Draw a group of pears that is 1 less. Write how
many you drew.

Name _____   Date _____

Color the group of 5 gray cubes. Then, count all the cubes in each tower and write how many. What do you notice?

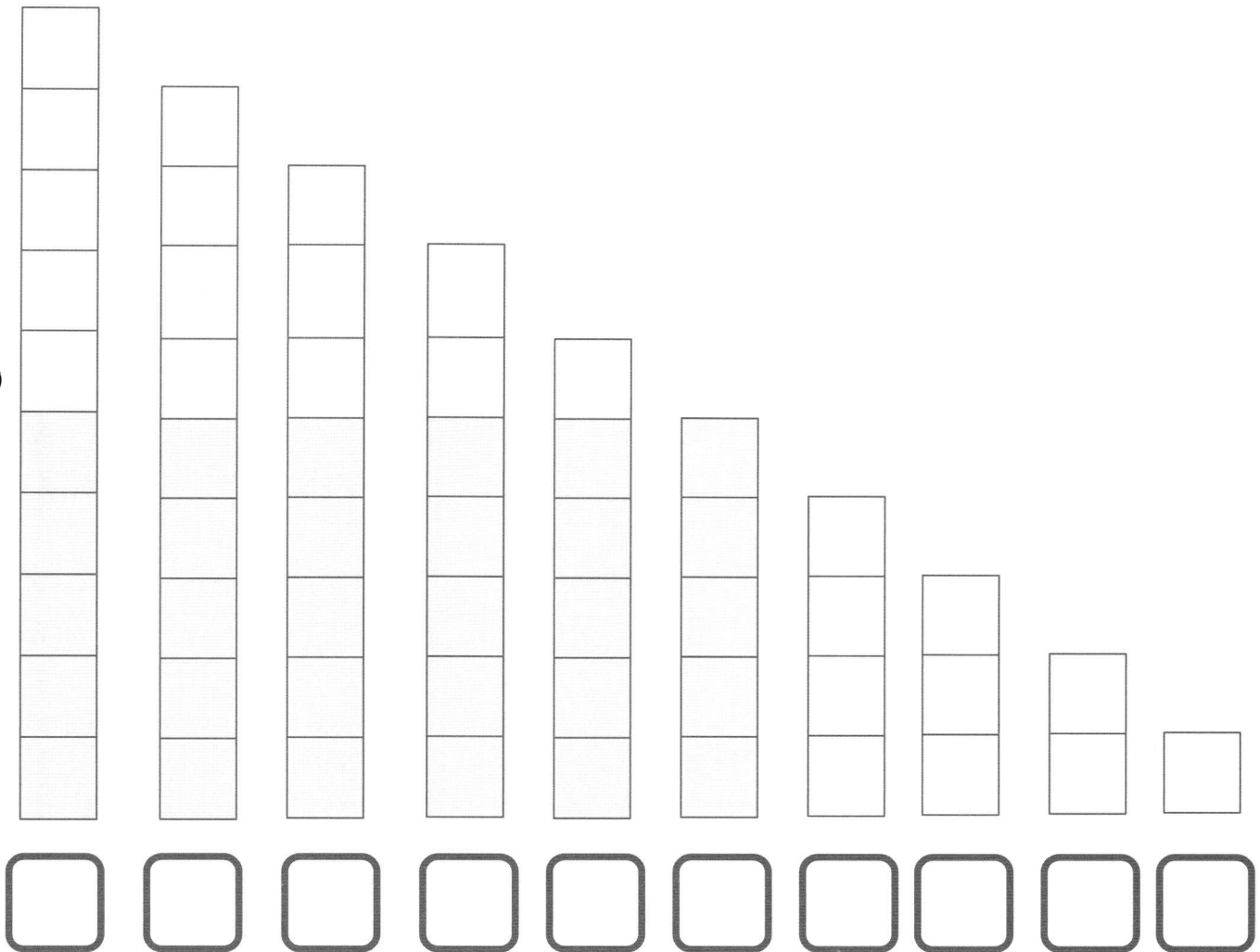

Count the number of cubes in a stair. Cross off the top cube. Use your words to say, "10. One less is nine. 9. One less is eight." Keep going all the way to the bottom of the stairs! Write how many cubes are in the stairs after you cross off the top cube.

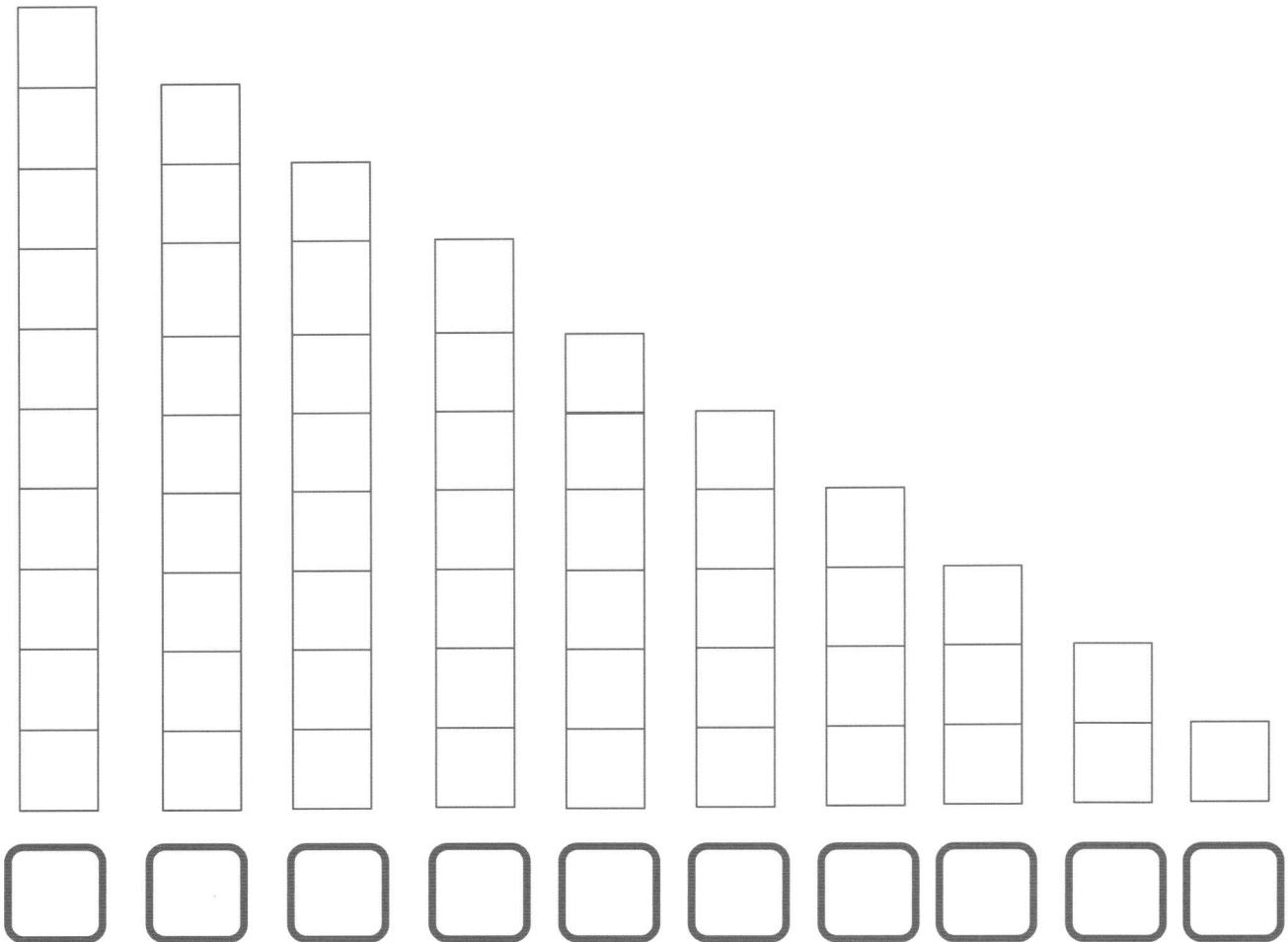

Name _____   Date _____

Count and say the number of cubes in the towers. Count the cubes that are crossed out. Say "1 less" and write the number.

Name _____   Date _____

Count and color the cubes in the tower. Cross the top cube off, and write the number. Draw the next tower with 1 less cube until there are no towers left.

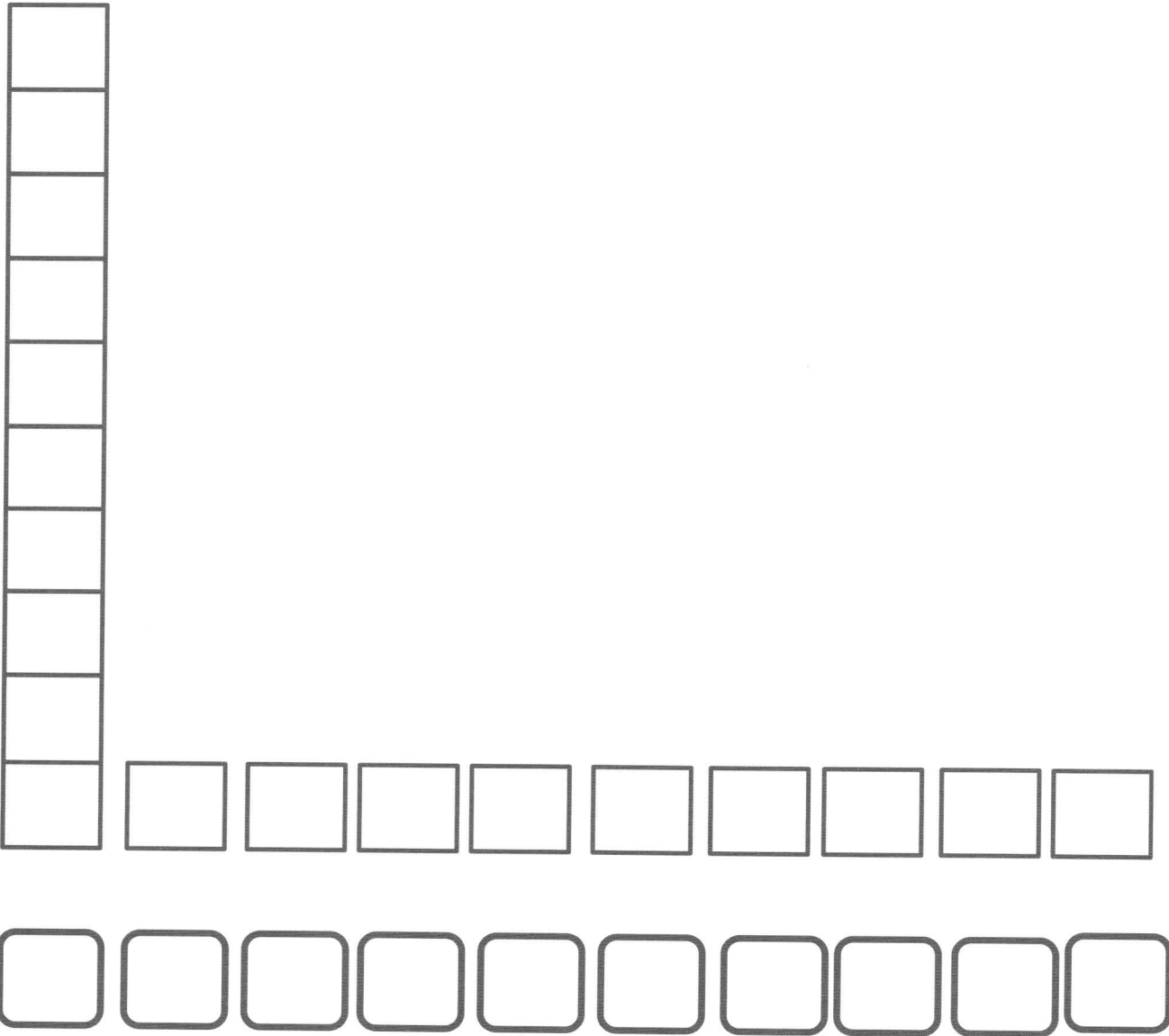

Name _____   Date _____

Count the objects. Write the number in the first box. Put an X on the shaded object. Count the objects that are left. Write the number that is left in the second box.

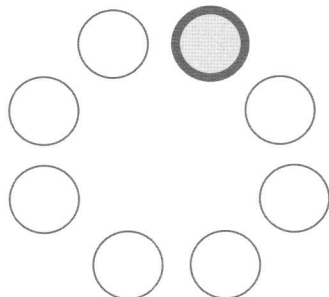

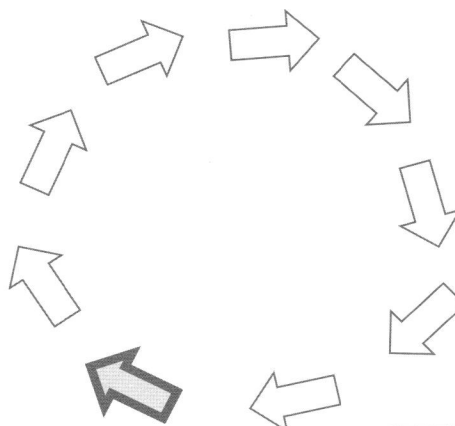

[   ] one less is [   ]   [   ] one less is [   ]

Count the objects. Write the number. Put an X on one object. Count the objects that are left. Write the number in the second box.

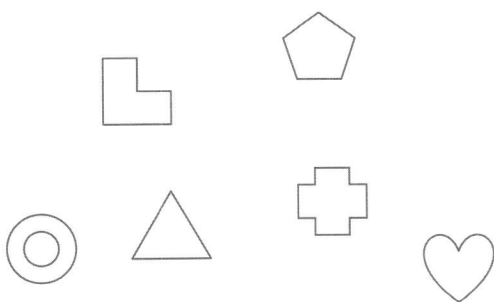

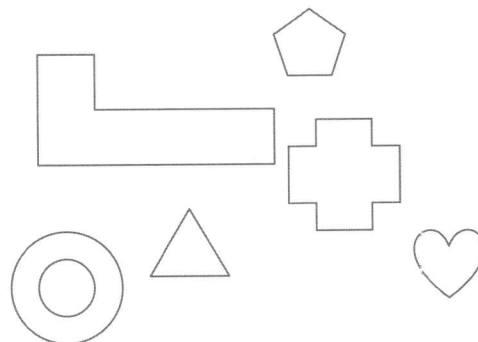

[   ] one less is [   ]   one less is [   ]

EUREKA MATH™

Lesson 36:   Arrange, analyze, and draw sequences of quantities that are 1 less in configurations other than towers.
Date: 6/6/14

161

Count and write how many.

Draw 1 less. Count and write how many.

Count and write how many.

Draw 1 less. Count and write how many.

EUREKA MATH

Lesson 36: Arrange, analyze, and draw sequences of quantities that are 1 less in configurations other than towers.

Date: 6/6/14

162

Name _____   Date _____

Fill in the missing numbers.

10, 9, ___, ___, ___, 5, 4, ___, ___, ___

Count and write the number of happy faces in the box. Draw another set
below it that has one less, and write the number in your set.

My set:

EUREKA
MATH™

Lesson 36:   Arrange, analyze, and draw sequences of quantities that are 1 less in
            configurations other than towers.
Date: 6/6/14

163

Name_____ Date_____

Draw bracelets to show 1 less than the number in the box.

If the number is missing, write it in the box.

| 10 | → | 9 | → | |

| 5 | → | 4 | → | |

Fill in the missing numbers.

_____, _____, 8, 7, _____, _____, _____, 3, 2, 1, _____

Name _____ Date _____

Count how many are in each group. Write the number in the box. Circle the smaller group.

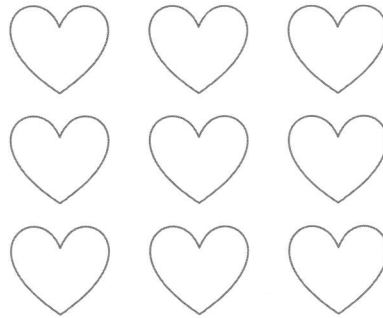

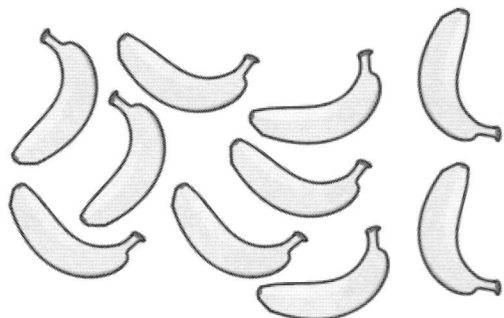

Draw some toys you enjoy.

How many? ☐

Draw some healthy foods.

How many? ☐

Student Name: _____

Topic A: Attributes of Two Related Objects

Rubric Score: _____ Time Elapsed: _____

	Date 1	Date 2	Date 3
Topic A			
Topic B			
Topic C			
Topic D			

Materials: (S) Module 1 assessment picture cards (cut out)

- T: (Identify the pictures as you place them in a row before the student.) Show me the pictures that are exactly the same.
- T: How are they exactly the same?
- T: Show me something that is *the same but* a little different.
- T: Use your words, "They are the same, but..." to tell me how the bears are different.

What did the student do?	What did the student say?

Topic B: Classify to Make Categories and Count

Rubric Score: _____ Time Elapsed: _____

Materials: (S) Module 1 assessment picture cards (cut out), sorting mat

   T: (Place all of the cards before the student.) Please sort the pictures into two groups on your sorting mat. (After sorting, have the student explain her reasoning.)

   T: (Point to the objects that went in the backpack.) Count the things that are in this group. (Look for the student to answer "3" rather than "1, 2, 3." If the student recounts to find the answer, ask again.)

Set the sort aside for the Topic D assessment.

What did the student do?	What did the student say?

Topic C: Numbers to 5 in Different Configurations, Math Drawings, and Expressions

Rubric Score: _____ Time Elapsed: _____

Materials: (S) 10 linking cubes

> T: (Put 5 loose cubes in front of the student.) Whisper-count as you put the cubes into a line. How many cubes are there?
>
> T: (Move the cubes into a circle.) How many cubes are there?
>
> T: (Scatter the cubes.) How many cubes are there?
>
> T: Please show this (show 2 + 1) using your cubes. (Have the student explain what she does. We might expect the student to make a linking cube stick of 3 and break it into two parts.)

What did the student do?	What did the student say?

Topic D: The Concept of Zero and Working with Numbers 0–5

Rubric Score: _____   Time Elapsed: _____

Materials: (S) Sort from Topic B (remove one identical bear for this assessment task so that there are 5 toys and 3 school items), numeral writing sheet

Note: Arrange the pictures as shown to the right. This arrangement is intended to give the student the opportunity to see 5 as *3 and some more, without recounting all.*

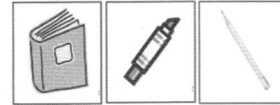

- T: How many things for school do you see? (Point to the top row.)
- T: (Point to the second row.) These are things we don't usually bring to school. How many are in this group? (Note if the student recounts all or determines the set of 5 *using the set of 3 in any way.*) How do you know it is 5?
- T: How many cats are shown here?
- T: Write your numbers in order from 0 to 5. (Note reversals, if any.)
- T: Write the number that tells how many toys there are.

What did the student do?	What did the student say?
Did the student show evidence of subitizing or recognizing embedded numbers, seeing 5 as 2 and 3 or 4 and 1?	

Module 1 Assessment Picture Cards

Sorting Mat

Student Name _____

Numeral Writing

Student Name: _____

	Date 1	Date 2	Date 3
Topic E			
Topic F			
Topic G			
Topic H			

Topic E: Working with Numbers 6–8 in Different Configurations

Rubric Score: _____ Time Elapsed: _____

Materials: (S) 10 linking cubes (or other familiar classroom object)

T: Please count 6 linking cubes, and put them in a row. (Pause.) Write the numeral 6.

T: (Arrange 7 cubes in a circular configuration.) Please count the cubes. (Pause.) Write the number 7. Show me the 5-group that's hiding in this group of cubes.

T: (Arrange 8 cubes into an array of 4 and 4.) How many cubes are there now? (Pause.) How did you know there were that many?

What did the student do?	What did the student say?
1.	
2.	
3.	

Topic F: Working with Numbers 9–10 in Different Configurations

Rubric Score: _____ Time Elapsed: _____

Materials: (S) 12 linking cubes (or other familiar classroom object), brown construction paper mat to show the problem

- T: Now, let's pretend these cubes are bears! Show me this problem: There were six bears who were eating leaves here in the woods. (Pause.) Three more bears came over to snack on some leaves. How many bears were eating leaves in the woods?
- T: Use your words to tell me how you figured out the problem.
- T: Write the number that tells how many bears there are eating leaves.
- T: Another bear came. Show me the bears now. How many bears is that? Write that number.

What did the student do?	What did the student say?
1.	
2.	
3.	
4.	

Topic G: *One More Than* **with Numbers 0–10**

Rubric Score: _____ Time Elapsed: _____

Materials: (T) 5-group cards (Lesson 7 Template, numeral side: 7, 8, and 9), 5-group card (Lesson 7 Template, dot side), 10 cubes

> T: (Hold up the card showing 4 dots.) Use the cubes to show me the number of cubes that is 1 more than this.

> T: (Hold up the card showing the numeral 7.) Use the number cards to show me the numeral that's 1 more. How did you learn that?

> T: Put these numeral cards in order from smallest to greatest. (Hand the students the 7, 8, and 9 cards out of order.)

What did the student do?	What did the student say?
1.	
2.	
3.	

Topic H: *One Less Than* with Numbers 0–10

Rubric Score: _____ Time Elapsed _____

Materials: (T) 5-group cards (Lesson 7 Template), 10 counting objects

- T: (Place 10 objects in an array of two 5-groups.) How many objects are there? (Note how the student counts.) Show 1 less. Write how many you have now.
- T: (Put the number cards in order from 10 to 1. Turn over the numbers 9, 7, 5, and 2.) Touch and tell me the hidden numbers. Don't turn over the cards, though!
- T: (Place the 9, 7, 5, and 2 dot cards in a line out of order.) Match the dot cards to the hidden numbers. Turn over the hidden card when you are sure you have matched it.

What did the student do?	What did the student say?
1.	
2.	
3.	

4 9

3 8

2 7

1 6

10 5

EUREKA MATH™

Module 1: Numbers to 10
Date: 6/20/14

Notes

Notes